This series aims to report new developments in physical research and teaching – quickly, informally, and at a high level. The type of material considered for publication includes:

1. Preliminary drafts of original papers and monographs

2. Lectures on a new field, or presenting a new angle on a classical field

3. collections of seminar papers

4. Reports of meetings

Texts which are out of print but still in demand may also be considered if they fall within these categories.

The timeliness of a manuscript is more important than its form, which may be unfinished or tentative. Thus, in some instances, proofs may be merely outlined and results presented which have been or will later be published elsewhere.

Publication of *Lecture Notes* is intended as a service to the international physical community, in that a commercial publisher, Springer-Verlag, can offer a wider distribution to documents which would otherwise have a restricted readership. Once published and copyrighted, they can be documented in the scientific libraries.

Manuscripts

Manuscripts are reproduced by a photographic process; they must therefore be typed with extreme care. Symbols not on the typewriter should be inserted by hand in indelible black ink. Corrections to the typescript should be made by sticking the amended text over the old one, or by obliterating errors with white correcting fluid. The figures (in the original size) ready for reproduction should be inserted into the text. Should the text, or any part of it, have to be retyped, the author will be reimbursed upon publication of the volume. Authors receive 50 free copies.

The typescript is reduced slightly in size during reproduction, therefore a large size of type should be used; best results will not be obtained unless the text on any one page is kept within the overall limit of 18 x 26.5 cm (7 x 10½ inches). The publishers will be pleased to supply on request special stationery with the typing area outlined.

Manuscripts in English, German or French should be sent to Springer-Verlag' 6900 Heidelberg, Postfach 1780.

Die ,, *Lecture Notes*" sollen rasch und informell, aber auf hohem Niveau, über neue Entwicklungen in der Physik berichten. Zur Veröffentlichung kommen:

1. Vorläufige Fassungen von Originalarbeiten und Monographien.

2. Spezielle Vorlesungen über ein neues Gebiet oder ein klassisches Gebiet in neuer Betrachtungsweise.

3. Seminarausarbeitungen.

4. Vorträge von Tagungen.

Ferner kommen auch ältere vergriffene spezielle Vorlesungen, Seminare und Berichte in Frage, wenn nach ihnen eine anhaltende Nachfrage besteht.

Die Beiträge dürfen im Interesse einer größeren Aktualität durchaus den Charakter des Unfertigen und Vorläufigen haben. Sie brauchen Beweise unter Umständen nur zu skizzieren und dürfen auch Ergebnisse enthalten, die in ähnlicher Form schon erschienen sind oder später erscheinen sollen.

Die Herausgabe der ,, *Lecture Notes*" Serie durch den Springer-Verlag stellt eine Dienstleistung an die physikalischen Institute dar, indem der Springer-Verlag für ausreichende Lagerhaltung sorgt und einen großen internationalen Kreis von Interessenten erfassen kann. Durch Anzeigen in Fachzeitschriften, Aufnahme in Kataloge und durch Anmeldung zum Copyright sowie durch die Versendung von Besprechungsexemplaren wird eine lückenlose Dokumentation in den wissenschaftlichen Bibliotheken ermöglicht.

Lecture Notes in Physics

Edited by J. Ehlers, München, K. Hepp, Zürich and
H. A. Weidenmüller, Heidelberg
Managing Editor: W. Beiglböck, Heidelberg

14

Methods of Local and Global Differential Geometry in General Relativity

Proceedings of the Regional Conference
on Relativity held at the University of
Pittsburgh, Pittsburgh, Pennsylvania,
July 13–17, 1970
Edited by D. Farnsworth, J. Fink, J. Porter
and A. Thompson

Springer-Verlag Berlin Heidelberg GmbH 1972

ISBN 978-3-540-05793-2   ISBN 978-3-540-37434-3 (eBook)
DOI 10.1007/978-3-540-37434-3

© by Springer-Verlag Berlin Heidelberg 1972.

Originally published by Springer-Verlag Berlin Heidelberg New York in 1972.

Library of Congress Catalog Card Number 72-75728.

Offsetdruck: Julius Beltz, Hemsbach/Bergstr.

FOREWARD

With the increased awareness among physicists and applied mathematicians
that global differential geometry and differential topology have an important
role to play in the further development of the arena of theoretical physics,
we feel it is important to inform young research physicists and mathematicians
of the problems to be formulated and solved in these applications of pure
mathematics. Consequently, a regional conference was organized and held
at the University of Pittsburgh, Pittsburgh, Pennsylvania, from July 13
through July 17, 1970. This volume contains some of the papers presented at
the Conference as well as notes on the lectures given by Professor Roger Penrose.
The manuscripts are published as supplied by the speakers, apart from changes
in format and obvious corrections.

We are grateful to the National Science Foundation for their support
of the conference (contract number GP 18904). We particularly wish to
thank Dr. Truman Botts of the Conference Board of the Mathematical Sciences
under whose auspices the Conference was organized. Thanks also go to
Miss Dolores Saylor, Mrs. Barbara Chamberlain, and Mr. Andrew R. Chopnak
for their assistance in the preparation of this volume.

TABLE OF CONTENTS

TECHNIQUES OF TOPOLOGY AND DIFFERENTIAL GEOMETRY IN GENERAL RELATIVITY[*]

David Lerner

University of Pittsburgh

Introduction

This paper presents some of the recently developed techniques for analyzing the global properties of space-time manifolds and uses them to prove the most recent of the "singularity theorems", due to S. W. Hawking and R. Penrose $\left(\left[5\right]\right)$. These theorems are rigorous mathematical proofs that under fairly general, physically acceptable hypotheses, weird things happen to our models of the universe. Examples of "weirdness" are unbounded curvature, causal geodesic incompleteness, closed timelike curves, and the like. For the most part, the singularity theorems do not assume that the metric tensor of space-time arises from a specific solution of the Einstein field equations. Thus they are immediately applicable to any theory in which the universe has the gross structure of a four-dimensional Lorentzian manifold.

In section 1, various properties of space-times and pseudo-Riemannian manifolds in general are reviewed. Timelike curves are shown to remain within the light cone (locally); and timelike geodesics are shown to be curves of maximal length (again, locally). Sections 2, 3 and 4 introduce some of the topological constructs which have proven very effective in studying the causal structure of space-time (see, for example $\left[1\right]$, $\left[2\right]$, $\left[4\right]$, and $\left[10\right]$). The space of causal curves is defined in section 5 and is shown to be a compact metric space under certain conditions. In the final section it is proven that, beyond a conjugate point, a causal geodesic is no longer a maximal curve. This, together with several earlier results, is sufficient to prove the singularity theorem.

---

[*]These notes, apart from a few modifications, are based on the series of lectures given by Professor R. Penrose at the conference.

There is very little heuristic discussion in this paper. The interested reader is referred to references $[4]$, $[10]$, and particularly to reference $[11]$ in which Professor Penrose elaborates considerably on all the topics covered here.

§ 1: BASIC DEFINITIONS; MATHEMATICAL PRELIMINARIES

1.1 A space-time (M, g) is a real four-dimensional differentiable manifold[*] without boundary on which a non-degenerate Lorentzian metric tensor g is globally defined. By Lorentzian is meant that for each x ∈ M there is a basis in T_x, the tangent space at x, relative to which the matrix of g_x has the form Diag $\{1, -1, -1, -1\}$.

1.2 Let $Y \in T_x$. Y is called spacelike if $g_x(Y,Y) < 0$, timelike if $g_x(Y,Y) > 0$, and null if $g_x(Y,Y) = 0$. The null cone at x is the set of all null vectors in T_x.

1.3 The space-time M (we will omit the "g" if there is no possibility of confusion) is said to be orientable with respect to time if there is a nowhere vanishing timelike vector field on M, say ξ . If this is the case, then for each x ∈ M, the component of timelike vectors containing ξ_x is called the set of future-directed timelike vectors at x. The timelike vectors in the other component are called past-directed. This choice having been made, M is said to be time-oriented. From now on, all space-times will be assumed to be time-oriented.

1.4 A path is a smooth map from a connected set in $\mathbb{R}$ to M.
A curve is the point-set image of a path. If $t \longrightarrow Y(t)$ is a path, the curve determined by $Y(t)$ will be denoted γ .

1.5 Let $Y(t)$ be a path with domain $D \subseteq \mathbb{R}$. Let a = inf D, b = sup D (possibly infinite). A point x ∈ M is said to be an endpoint of $Y(t)$ if for all

---

[*]The words "differentiable" and "smooth" are taken to mean C^∞ .

The map is assumed to have no critical points; i.e. to have a non-zero tangent vector at all points.

sequences $\{t_i\} \subseteq D$ such that $t_i \rightarrow a$, $\gamma(t_i) \rightarrow x$. [Similarly for $t_i \rightarrow b$.] If x is an end point of the path $\gamma(t)$, it is also an end point of every path in the curve determined by $\gamma(t)$. Thus it makes sense to speak of the end point of the curve γ.

1.6 The Exponential Map: Denote by ∇ the unique symmetric affine connection on M satisfying $\nabla g = 0$. Let U be a patch on M with coordinates (u^i). The Christoffel symbols are defined in U by $\nabla_{\frac{\partial}{\partial u^i}}\left(\frac{\partial}{\partial u^j}\right) = \Gamma^k_{ij}\frac{\partial}{\partial u^k}$. Suppose $\gamma(t)$ is a path $t \rightarrow u^i(t)$. By definition, $\gamma(t)$ is a geodesic iff:

$$\frac{d^2 u^k}{dt^2} + \Gamma^k_{ij}\frac{du^i}{dt}\frac{du^j}{dt} = 0.$$ Conversely, given this system, the

existence and uniqueness theorem for ordinary differential equations gives the immediate result: For each $p \in$ M there is a neighborhood $W(p)$, and for each $x \in W(p)$ a neighborhood V_x of the origin in T_x such that

a: For $v \in V_x$, there is a unique geodesic $\gamma_v(t)$ defined for $|t| < \varepsilon(p)$ satisfying $\gamma_v(0) = x$ and $\frac{d\gamma_v}{dt}(0) = v$

b: The subset of V_x consisting of those v such that 1 is in the domain of γ_v contains a starlike open neighborhood of 0, $V'(x)$. (This makes sense: if $v \in V_x$, then $\gamma_v(t)$ is defined for $|t| < \varepsilon(p)$; choose $r < \varepsilon(p)$ Then $\gamma_v(rt)$ is still a geodesic and is defined for $|t| \leq 1$. Since $\frac{d\gamma_v(rt)}{dt}(0) = r\frac{d\gamma_v}{dt} = rv$, by uniqueness it follows that $\gamma_v(rt) = \gamma_{rv}(t)$. Thus $\gamma_w(t)$ is defined for all $w \in T_x$ such that $w = sv$ where $|s| \leq r$.)

The exponential map is defined on V'_x by $\exp_x(v) \equiv \gamma_v(1)$. Thus the geodesic $\gamma_v(t)$ is described by $t \rightarrow \exp_x(tv) = \gamma_v(t)$. It can be shown ([8]) that $\exp_x$ is non-singular at the origin - thus there is a neighborhood $V''_x \subseteq V'_x$ on which $\exp_x$ is a diffeomorphism. The open set $\exp_x(V''_x)$ is called a normal neighborhood of x. If one takes for coordinates at x the images under $\exp_x$ of Cartesian coordinate functions in T_x, these are called normal coordinates. It can further be shown ([6]) that there is a basis for the topology of M consisting of <u>geodesically convex</u> sets (U is geodesically convex if any two points of U may be joined by a unique geodesic lying entirely in U). The basis may be further refined to consist of geodesically convex sets with compact closures, whose boundaries are diffeomorphic to S^3. Such sets will be called <u>simple regions</u> and will be used extensively throughout these notes. Clearly, if N is a simple region and $x \in N$, $\exp_x^{-1}$ is a diffeomorphism on N .

1.7 <u>Families of geodesics; Jacobi Fields</u> ([6] , [8])

Let $B = \left\{ (x,t) : |x| < \varepsilon , a < t < b \right\}$ be an open rectangle in $\mathbb{R}^2$. A smooth map $f : B \longrightarrow M$ is called a parametrized surface in M. By a <u>vector field Q on f</u> is meant a function which assigns to each $(t,x) \in B$ a tangent vector $Q(x,t) \in T_{f(x,t)}$. The vector fields $f_* \left(\frac{\partial}{\partial t} \right)$ and $f_* \left(\frac{\partial}{\partial x} \right)$ will be denoted by $T(x,t)$ and $X(x,t)$ respectively. The map f need not be one-to-one, so the vector fields T and X will <u>not</u>, in general, be the restrictions to f(B) of vector fields on M.

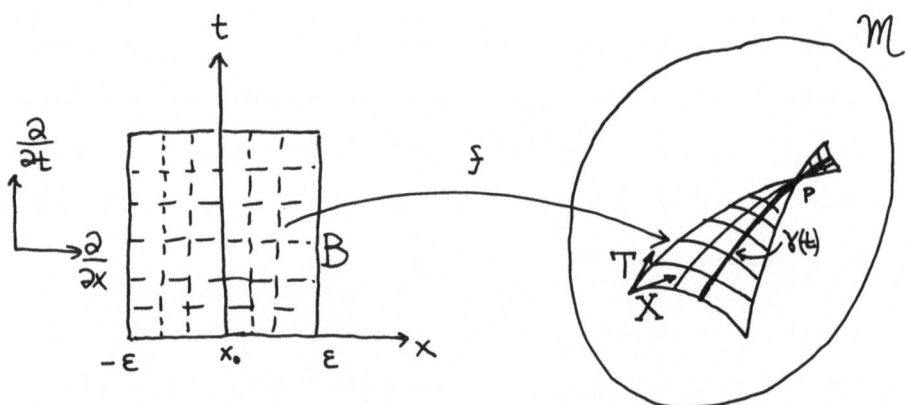

fig. 1: f must be smooth but not necessarily 1-1. At p, the vector field T on
f fails to be the restriction to f(B) of any vector field on M. X
"connects" corresponding points on neighboring X = constant curves.
In the particular case that f maps the X = constant curves to geodesics in
M affinely parametrized by t, f is called a <u>one-parameter family of geodesics</u>,

and X is called the <u>variation vector field</u>. If γ(t) is the geodesic given by $t \longrightarrow f(x_0,t)$, then the vector field on γ defined by $t \longrightarrow X(x_0,t)$ is called a <u>Jacobi Field</u> along γ. Intuitively, a Jacobi field may be thought of as a field of connecting vectors, joining the points $f(x_0,t)$ and $f(x_0+\delta, t)$ on neighboring geodesics.

<u>Lemma 1</u>: For any Jacobi field X on γ(t), $\dfrac{D^2 X}{dt^2}$ = R(T,X)T, where R is the curvature tensor of M.

> <u>Proof</u>: Let γ(t) = $f(x_0,t)$. Then the assignment $t \longrightarrow X(x_0,t)$ is one-to
> one, as is the assignment $t \longrightarrow T(x_0,t)$, since t is an affine
> parameter and cannot be stationary. Thus ∇_T and ∇_X are
> meaningful at points of γ(t). Since X is Lie propagated along
> γ(t), $\nabla_X T = \nabla_T X$ on γ. We have:
>
> $$R(T,X)T = \nabla_T \nabla_X T - \nabla_X \nabla_T T - \nabla_{[T,X]} T$$
> $$= \nabla_T \nabla_X T = \nabla_T \nabla_T X = \frac{D^2 X}{dt^2} .$$

This is the familiar Jacobi equation, or equation of geodesic deviation, and in fact a Jacobi field is often defined to be a vector field along a geodesic which satisfies this equation. However, the two definitions are equivalent, since any solution to the deviation equation is the variation vector field, restricted to γ(t), of a 1 - parameter family of geodesics. The solutions to the Jacobi equation along γ(t) form an eight-dimensional vector space. Any such solution is uniquely determined by prescribing X and DX/dt at some point. Inside a simple region, one has the further result that X(t) is uniquely determined by its values at the end points of γ.

1.8 Since tangent vectors to geodesics are parallelly propagated, it makes sense to speak of null, timelike and spacelike geodesics. At any point x ∈ M, the <u>light cone</u> at x, denoted $\mathcal{L}_x$ is defined by $\mathcal{L}_x = \{p : p$ lies on a null geodesic through $x\}$. A smooth three-dimensional submanifold of M is called a <u>null</u> (timelike, spacelike) <u>hypersurface</u> if the surface normal is null (spacelike, timelike) at each point.

<u>Lemma 2</u>: Let $x \in N$, a simple region. Then $\mathcal{L}_x \cap N$ is a null hypersurface (except at x).

<u>Proof</u>: Let X_p be tangent to the surface at p. There is a unique $T \in T_x$ such that $\exp_x(T) = p$. T is a null vector at p and is tangent to the null geodesic $t \longrightarrow \exp_x(tT) \equiv \gamma(t)$. Claim $g_p(T,X) = 0$: By the above remarks, there is a unique Jacobi field X(t) defined on $\gamma(t)$ satisfying X(0) = 0, X(1) = X_p.

$$\frac{d}{dt}\left[g(X,T)\right] = g(\nabla_T X, T) + g(X, \nabla_T T)$$
$$= g(\nabla_T X, T) = g(\nabla_X T, T)$$
$$= \tfrac{1}{2}\nabla_X\left[g(T,T)\right] = 0, \text{ since } g(T,T) = 0.$$

Thus g(X,T) is independent of t on γ . But at t = 0, X = 0 and $g_x(X,T) = 0$. So g(T,X) = 0 on γ , and at p in particular. Thus T is the surface normal at p.

<u>Lemma 3</u>: For any $x \in N$, a simple region, the timelike (spacelike) geodesics through X are the orthogonal trajectories of the hypersurfaces
$$S_{t_0}(x) \equiv \left\{ y: y = \exp_x(v), \; g_x(v,v) = t_0 \right\} \cap N, \; t_0 > 0 \; (t_0 < 0).$$

<u>Proof</u>: Let γ_p be tangent to $S_{t_0}(x)$ at p. Let $u \longrightarrow w(u)$ be an integral curve of γ_p lying in $S_{t_0}(x)$. We may write $w(u) = \exp_x\left[V(u)\right]$ where $g_x(V(u), V(u)) = t_0$, and $u \longrightarrow V(u)$ is a smooth curve in the domain of $\exp_x$. The map $f(t,u) = \exp_x\left[t \cdot V(u)\right]$ is a one-parameter family of geodesics. The integral curves of the variation vector field lie in the hypersurfaces $S_t(x)$. Let $\gamma(t) = f(t,u_0)$ be the radial geodesic striking p. Along γ we have

$$\nabla_T \left[g(T, U) \right] = g(\nabla_T T, U) + g(T, \nabla_T U)$$
$$= \frac{1}{2} \nabla_u \left[g(T, T) \right] = 0 \quad \text{since } g(T, T) = t_o.$$

Thus, as before, $g(T, U)$ is independent of t on γ . At t = 0,

$f(o,u) = \exp_x(0) = 0$; so $U(o, u_o) = 0$. Thus $g(T, U)$ is identically

zero on γ (and on all the radial geodesics, for that matter). In

particular, $Y_p = U_p \Leftarrow g_p(T, Y) = 0$ as asserted.

1.9 Suppose now that $\omega : [a, b] \to N$ is a smooth future-directed path lying

inside $\mathcal{L}_x$ and to the future of $S_{t_o}(x)$. For each $u \in [a, b]$, let $V(u)$

be the point on $S_{t_o}(x)$ determined by the unique timelike geodesic joining

x and $\omega(u)$. Then there is a unique unit vector $V(u)$ satisfying

$\exp_x(t_o V(u)) = V(u)$. The parametrized surface given by $(t, u) \xrightarrow{f} \exp_x \left[t \, V(u) \right]$

contains the curve $\omega(u)$. We may write $\omega(u) = (t(u), u)$, and in

terms of the vector fields U and T ($f_* \frac{\partial}{\partial u} + f_* \frac{\partial}{\partial t}$), $\frac{d\omega}{du} = t'(u) T + U$.

Define the length of ω by

$$L(\omega) \equiv \int_{u=a}^{u=b} \sqrt{g \left(\frac{d\omega}{du}, \frac{d\omega}{du} \right)} \, du$$

Lemma 4: With the above notation, $L(\omega) \leq t(b) - t(a)$ with equality holding iff.

$V(u)$ is constant - i.e. iff γ is a geodesic curve.

Proof: $g \left(\frac{d\omega}{du}, \frac{d\omega}{du} \right) = \left[t'(u) \right]^2 \cdot g(T, T) + 2t'(u) g(T, U) + g(U, U)$

$= \left[t'(u) \right]^2 + g(U, U)$, since $g(T, T) = 1$ and

$g(T, U) = 0$ as in Lemma 3.

$\leq \left[t'(u) \right]^2$, since $g(U, U)$ must either be

spacelike or zero.

Hence $\int_{u=a}^{u=b} \sqrt{g(\frac{d\omega}{du}, \frac{d\omega}{du})} \, du = L(\omega) \leq \int_a^b |t'(u)| \, du = t(b) - t(a)$.[*]

Thus the length function is maximized locally along timelike geodesics.

---

[*]That t(u) is monotone increasing follows from the requirement that a path
have a non critical parameter, and that $\omega(u)$ is timelike and future-directed.

§2: CAUSALITY AND CHRONOLOGY

From this point on, the word "curve" will be taken to mean a curve containing its endpoints ⟨if they exist; for example, the curve γ in fig. 2 has no future endpoint⟩. The same applies to paths.

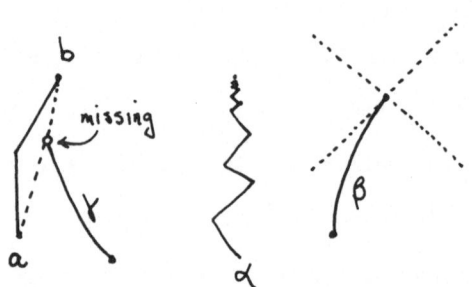

fig.2: $\mathcal{M}^2 \setminus \{point\}$. γ is a curve without a future endpoint.
α is not a trip; there is no well-defined future-pointing tangent at the future endpoint.
β is not a timelike curve, because the tangent vector at its future endpoint is null.

2.1 A <u>trip</u> is a future-directed piecewise timelike geodesic. Degenerate geodesics are not allowed. A trip with past endpoint x and future endpoint y is called a trip from x to y. The set of points at which a trip fails to be smooth has no limit point; this follows from the requirement that a trip contain its endpoints (see fig. 2.). If there is a trip from x to y we write x < < y which is read "x chronologically precedes y".

2.2 The set $I^{+}(x) \equiv \{y; x \ll y\}$ is called the <u>chronological future</u> of x.

   $I^{-}(x) \equiv \{y: y \ll x\}$ is called the <u>chronological past</u> of x.

<u>Remark</u>:

   Note that an event x is in its own future (x≪x) iff. there is a closed trip containing x. Also note that is is necessary to include "piecewise" in the definition of trip; in fig. 2, the points a and b may be joined by a timelike curve, but not be a single timelike geodesic, and we clearly want b ε $I^{+}(a)$. In fact, one may easily construct examples in which an arbitrarily large number of "joints" are needed to construct a trip connecting two points which lie on a smooth timelike curve.

Finally, one does not wish to ignore the "non-existence" of the future endpoint of γ. This is detectable by other means ($\S$ 5).

2.3. <u>Proposition</u>: Trips and smooth future-directed timelike paths are equivalent in the study of causality in spacetimes. More precisely:

x<<y <=> there is a smooth future-directed timelike path from x to y.

<u>Proof</u>: Let α be a trip from x to y. It must be shown that the joints on α may be smoothed. If α is not smooth at A, there is a simple region N containing A and no other joints. Let $\ell_1 \, \& \, \ell_2$ be the timelike geodesics meeting at A, and let L_1 and L_2 denote their images in T_A by $\exp_A^{-1}$. L_1 and L_2 are rays lying inside the null cone and meeting at the origin. Choose one of these, L_2 say, to be the time axis in a Minkowskian coordinate system (t,x,y,z). If $r^2 = x^2 + y^2 + z^2$ then the locus given by $t^2 = \lambda r^2, \; \lambda > 1$ is a cone whose generators are timelike lines. For some λ_0, L_1 lies on $t^2 = \lambda_0 r^2$. Choose λ^* so that $\lambda_0 > \lambda^* > 1$. Then the cone $t^2 = \lambda^* r^2$ contains both L_1 and L_2 in its interior. Construct a copy of this cone $\langle$ by translating from the origin $\rangle$ at every point in $\exp_A^{-1}$ (N). Since the inverse image of the light cone at A properly contains the λ^* cone, and since the light cones in N are a smoothly varying family of hypersurfaces, there must be an open nbd. of A, $U \subseteq N$, such that the inverse image of the light cone at any $p \in U$ contains the cone at $\exp_A^{-1}$ (p) - (see fig. 3 a.)

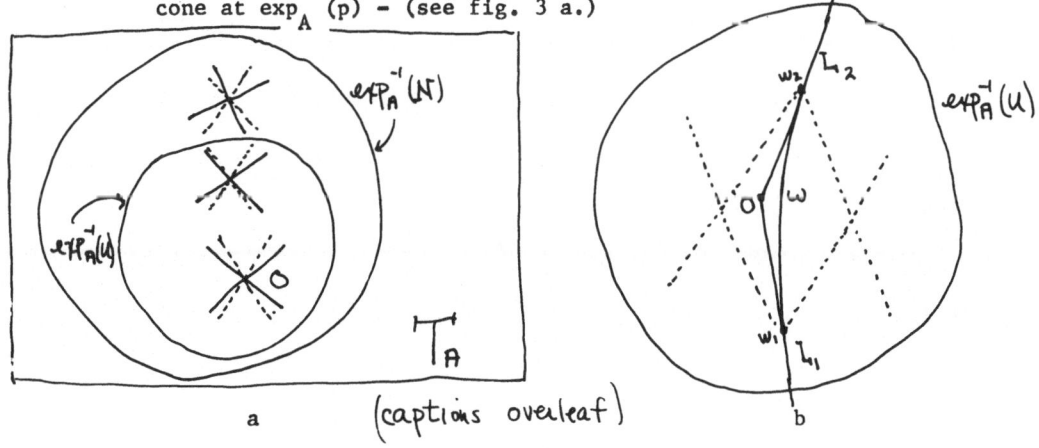

a   (captions overleaf)   b

fig. 3 a: The dotted lines represent the λ^* cones. The solid lines represent the inverse images of light cones under $\exp_A^{-1}$.

fig. 3 b: w joins w_1 and w_2 and "smoothes the joint". The tangent vector to w at any pt. lies within the λ^* cone at that point.

Now choose w_2 on L_2, w_1 on L_1, both in $\exp_A^{-1}$ (U). Since the line $\begin{bmatrix} L_1 & L_2 \end{bmatrix}$ lies within the λ^* cone, it is a fact from analysis that the joint may be replaced by a smooth curve w whose tangent vector at each point lies in the λ^* cone at that point and which joins L_1 and L_2 smoothly at w_1 and w_2. By construction, $\exp_A(w)$ is then timelike and the first part of the proof is completed.

Conversely, let $\gamma(t)$ be a smooth timelike path from x to y. Let $q = \gamma(t_o)$, and consider $\exp_q^{-1}[\gamma(t)]$. Since $d\gamma/dt(t_o)$ is timelike and smooth there is a neighborhood V(q) of the origin in T_q such that $\exp^{-1}[\gamma(t)] \cap V(q)$ lies in the null cone. Set $W(q) \equiv \exp_q[V(q)]$. Since γ is compact, it is covered by finitely many of these sets, $\{W(q_i) : 1 \le i \le \kappa\}$ where $q_i \ll q_{i+1}$. Choose points $y_1 = x$, y_i on $\gamma \cap W(q_i) \cap W(q_{i+1})$, $y_{\kappa+1} = y$. By construction $x = y_i$ may be joined to q by a timelike geodesic, q_1 to y_2, and so on. Thus we have a trip from x to y.

2.4 A <u>causal trip</u> is a future-directed piecewise geodesic path whose pieces may be either timelike or null. Constant paths are considered degenerate null geodesics, so every point of M is connected to itself by a causal trip. If there is a causal trip from a to b, we write a < b; and a is said to <u>causally precede</u> b. Clearly a<<b => a < b.

2.5 The <u>causal future</u> of a is defined by $J^+(a) \equiv \{x : a < x\}$. $J^-(a) \equiv \{x : x < a\}$ is called the <u>causal past</u> of a. We have $I^\pm(a) \subseteq J^\pm(a)$.

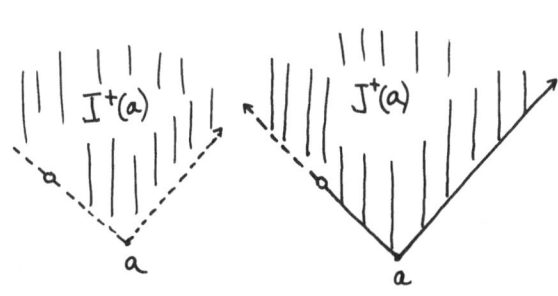

fig.4: In M^4, $J^+(o) = \{(t,x,$ $y,z):$ $t \geqslant o$, $t^2 \geqslant x^2$ $+ y^2 + z^2\}$. Here $J^+(x)$ $= \overline{I^+(x)}$, but this is not the case in general: it is not true in $M^2 \smallsetminus \{p\}$.

2.6 <u>Proposition</u>: $I^+(a)$ is open, for all $a \in M$. Similarly for $I^-(a)$.

    <u>Proof</u>: Let $b \in I^+(a)$, N a simple region containing b. Let γ be a trip from a to b, and let ℓ_n be the terminal segment of this trip. Choose $\mathfrak{z}$ on ℓ_n such that $\mathfrak{z} \in N$ and $\mathfrak{z} \ll b$. $\exp_{\mathfrak{z}}^{-1}(N)$ is an open nbd. of the origin in $T_{\mathfrak{z}}$. Let J be $\exp_{\mathfrak{z}}^{-1}(N)$ intersected with the interior of the future null cone. J is open and contains $\exp_{\mathfrak{z}}^{-1}(b)$. Thus $\exp_{\mathfrak{z}}(J)$ is open, contains b, and has the property that any of its points may be joined to $\mathfrak{z}$ by a timelike geodesic, so $\exp_{\mathfrak{z}}(J) \subseteq I^+(a)$.

2.7 <u>Proposition</u>: Let N be any simple region, and let $\gamma(t)$ be any piecewise smooth timelike path which intersects N. Let $b = \gamma(t_0)$ be an arbitrary point in $N \cap \gamma(t)$. Then $\gamma(t) \cap N$ lies within the light cone at b. $\langle$ It is locally true that a timelike curve cannot escape from the light cone.$\rangle$ *

    <u>Proof</u>: Suppose not. Since $\gamma(t)$ is piecewise smooth, it must lie within the light cone for some finite distance; let $\gamma(t_1) = c$ be the first point of intersection with the light cone at b. For $t \in (t_0, t_1)$, let R_t be the radial timelike geodesic from b to $\gamma(t)$; and let γ_t denote the portion of γ from t_0 to t. By Lemma 5, and since γ is

---

*In many curved space-times, this result does not hold outside of simple regions - see § 6.

timelike, we have $0 < L(\gamma_t) \leqq L(R_t)$. Since γ is continuous, both $L(\gamma_t)$ and $L(R_t)$ are continuous functions of t; further more, $L(\gamma_t)$ is a monotone increasing function of t. Thus at $t = t_1$, we must have $0 < L(\gamma_{t_1}) \leq L(R_{t_1}) = 0$, which is impossible.

2.8 <u>Proposition</u>: $a<<b, \ b < c \Rightarrow a<<c$

$a < b, \ b<<c \Rightarrow a<<c$

<u>Proof</u>: The proof clearly reduces to showing that if a and b are joined by a timelike geodesic, b and c by a null geodesic, then there is a trip from a to c.

Cover $[bc]$ by finitely many simple sets $N_1, --, N_K$. Let x_i be the future intersection of $[bc]$ with ∂N_i (see fig 5). Now b is on the

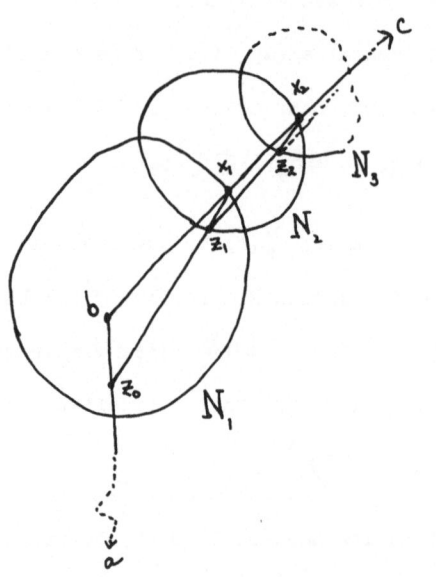

past light cone of x_1 and hence there is z_0 on $[ab] \cap N_1$ such that $z_0 \ \epsilon \ I^-(x_1)$. This is true in the simple region N_1 ⟨we are making use of the fact that the proposition is clearly true in Minkowski space⟩. Then $[z_0 \ x_1]$ is a timelike geodesic which intersects ∂N_2 at some pt. z_1. Now z_1 is in the past of x_2 and thus there is a timelike geodesic $[z_1 x_2]$ intersecting ∂N_3 at z_2. Etc.

Thus $[a_0 z_0 z_1 \cdots z_{k-1} c]$ is the required trip.

<u>Remark</u>: It is evident that $I^-(a)$ is open for all a, and that $a < b, \ b<<c \Rightarrow$ $a<<c$ - these statements are the so - called "time reverses" of statements already proven and are obtainable from them by replacing ⟨ in the proof⟩ the word "future" by "past", "precedes" by "follows", etc. We will have occasion to use the time-reverses of many propositions and

theorems.

2.9　The curve γ is a <u>causal</u> <u>curve</u> <=> $\forall a, b \in \gamma$, and all neighborhoods Q of the segment $[ab]$, there is a causal trip from a to b lying in Q. γ is continuous, but not necessarily smooth. It is not too difficult to see that in a simple region, if one pulls a causal curve back into the tangent space with the exponential map, that the definition forces $\exp^{-1} (\gamma)$ to be a curve of bounded variation with respect to the standard norm on R^4. Thus γ is smooth almost everywhere. The essential idea is that the causal curves form the complete set of pointwise limits of the causal trips. This will be covered in detail in section 5.

A causal trip clearly determines a causal curve. Conversely, if there is a causal curve from a to b, there is a causal trip from a to b.

2.10　<u>Remark</u>:　If a < b, but a<<b is false, there is a null geodesic from a to b. The converse is false:

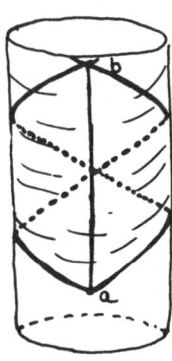

fig.6:　A portion of a two-dimensional Einstein universe. The future of a is shaded. b lies on a null geodesic containing a, but there is also a timelike geodesic from a to b. The point b lies outside the region in which $\exp_a$ is a diffeomorphism.

2.11　Let $S \subseteq M$. Then we define $I^+[S]$, the <u>future of S</u> by $I^+[S] = \bigcup_{x \in S} \{I^+(x)\}$. $I^-[S]$ is defined similarly. By 2.6, $I^+[S]$ is open for any set S.

2.12　<u>Proposition</u>:　(a):　$\overline{I^+[S]} = \{x: \ I^+(x) \subseteq S\}$

　　　　　　　　　　(b):　$\partial I^+[S] = \{x: \ I^+(x) \subseteq S \text{ and } x \notin S\}$

<u>Proof</u>:　(a):　If $I^+(x) \subseteq S$, $x \in \overline{I^+(x)} \subseteq \overline{I^+ S}$. Conversely, if $x \in \overline{I^+[S]}$, let $y \in I^+(x)$. Since $I^-(y)$ is open and contains x, there is an open

nbd. $U(x) \subseteq I^-(y)$. Since $U(x) \cap I^+[S] \neq \phi$, y is in the future

of some element of S. $\therefore I^+(x) \subseteq I^+ S$.

(b): Follows from (a) and the fact that $I^+[S]$ is open.

2.13 Let $F \subseteq M$. If there is a set S such that $F = I^+[S]$, then F is called a <u>future</u>

<u>set</u>.

2.14 <u>Proposition</u>: F is a future set $\iff I^+[F] = F$

<u>Proof</u>: If $F = I^+[F]$ then F is a future set by definition. Conversely, if F

is a future set, $\exists S \subseteq M$ such that $F = I^+[S]$. Then $I^+[F] =$

$I^+[I^+[S]] = I^+[S] = F$.

2.15 <u>Proposition</u>: Let $Q \subseteq M$. The following are equivalent:

(a) $I^+[Q] \subseteq Q$

(b) $Q^0 = I^+[Q]$

(c) $I^+[Q] \cap I^-[\sim Q] = \phi$

(d) $I^-[\sim Q] \subset \sim Q$

<u>Proof</u>: exercise. If Q satisfies the above, it is impossible to get out of

Q on a timelike path.

2.16 A subset S of M is called <u>achronal</u> iff. no two points of S are

chronologically related (iff. $\forall x,y \in S$, neither x<<y nor y<<x)

A subset B of M is called an <u>achronal boundary</u> iff. there exists a subset S

of M such that $B = \partial I^+[S]$

<u>Remark</u>: A set may be locally spacelike and fail to be achronal - for example, in

M^3 take a right circular helix with pitch less than $\pi/4$:

The set t = 0 in M^4 is achronal, and so is any subset

of this set.

The set $t^2 = x^2 + y^2 + z^2$, $t \geqslant 0$ is also achronal.

(Null surfaces can be achronal.)

2.17 <u>Proposition</u>: B, an achronal boundary, is an achronal set.

  <u>Proof</u>: $B = \partial I^+[S]$, some S. If x, y ε B satisfy x<<y, then y $\epsilon I^+(x) \subseteq I^+[S]$ which is open. Hence y cannot be on the boundary of $I^+[S]$.

2.18 If $B = \partial F$ is an achronal boundary of the future set F, put $P = \sim \overline{F}$. P is open and is a past set in the sense that $I^-[P] = P$ (exercise). Clearly $B = \partial P$ also and so we have the

  <u>Proposition</u>: A non-empty achronal boundary separates M into two disjoint open sets F and P, where F is a future set and P is a past set. If a ε P and b ε F and a<<b, any trip from a to b intersects B in precisely one point

  <u>Proof</u>: The last statement is the only one not proven. But the trip is a connected set lying partly in P and partly in F; so it must intersect B. That it intersect B only once follows from the fact that B is an achronal set.

2.19 <u>Remark</u>: In M^4, for any point a, if $B = \partial I^+(a)$, then $M^4 = B \cup I^+[B] \cup I^-[B]$
    This is false in general. (fig. 7)

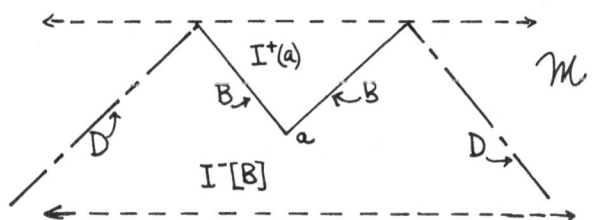

fig. 7: Here M is just a spacelike open strip in M^2. $\mathcal{M} \neq B \cup I^+[B] \cup I^-[B]$; also, B is not a maximal achronal set – $B \cup D$ is also achronal. A given spacetime need not have any achronal boundaries— eg. identify the upper and lower edges of the figure; then $I^+(a) = M$ for every a.

In view of prop. 2.18, one would suspect that achronal boundaries, when they do exist, are reasonable topological objects. In fact, we have the following:

2.19 <u>Proposition</u>: Any achronal boundary is a topological (C^{o}) 3 - manifold.

    <u>Proof</u>: We only sketch the proof. It is necessary to show that each b ε B has a neighborhood ⟨in the relative topology induced by M⟩ homeomorphic to an open ball in $\mathbb{R}^{3}$. Introduce normal coordinates (t,x,y,z) at b in such a way that each curve x,y,z = constant is a timelike curve from P to F. By prop 2.18, each such curve intersects B in a unique point, which is then mapped to the point (x,y,z) ε $\mathbb{R}^{3}$.

2.20 A causal geodesic segment is called <u>past endless</u> (<u>future endless</u>) if it has no past (future) endpoint. Remember that curves and paths are always <u>assumed</u> to contain their endpoints if they exist. So to say that a geodesic is past endless is to say that it has been prolonged to the past as far as possible. A geodesic which is both past and future endless is called simply, <u>endless</u>. Clearly <u>all</u> geodesics are endless. The use that will be made of the term "endless" is indicated in fig. 8.

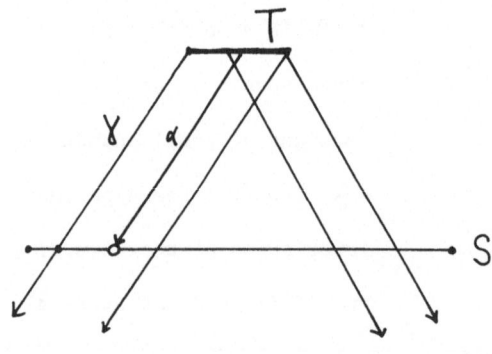

fig. 8

The statement "Any past-directed null geodesic with future endpoint on T either has a past endpoint on S or is past endless." is to be interpreted as follows:

    All the geodesics in question are to be prolonged to the past as far as possible. Those, like γ , which intersect S in the process are said to have past end points on S. The geodesic α , however, is past endless without intersecting S.

2.21 Examples:

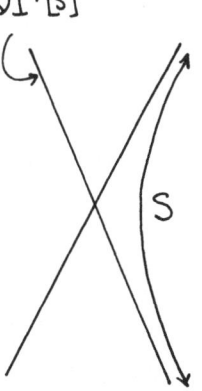

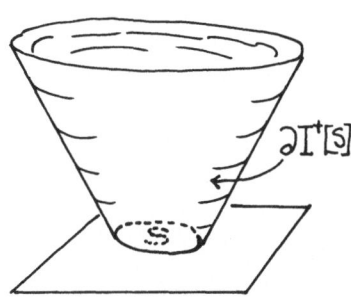

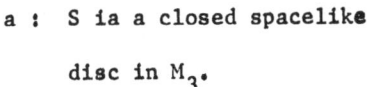

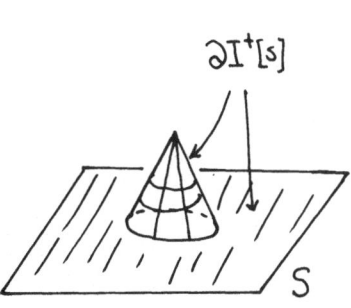

a : S ia a closed spacelike   b : S is a spacelike   c : S is a uniform

    disc in M_3.   plane in M^3 with   acceleration

                               disc removed.   hyperbola. $\partial I^+[S]$

                               $\partial I^+[S]$ is S together   is a null hyper-

                               with the cone.   plane.

2.22 <u>Proposition</u>: Let B = ∂F, where F is a future set. Let x ϵ B and suppose

there is an open set Q containing x satisfying:

      (a): If y ϵ Q $\cap$ F, there is a point z in F $\smallsetminus$ Q, such that $z <\!< y$.

    or equivalently

      (b): $I^-(y) \cap F \smallsetminus Q \neq \phi, \; \forall y \in Q \cap F$.

    or equivalently

      (c): $F = I^+[F \smallsetminus Q]$

Then x is the future endpoint of a null geodesic lying in B.

<u>Proof</u>: (a) <=> (b) <=> (c) (exercise). Let N be a simple region such that

    $x \in N \subseteq \overline{N} \subseteq Q$.   Choose $\{y_i\} \subseteq N \cap F$ such that $y_i \to x$; the y_i

    are $(by (a))$ future endpoints of trips to $F \smallsetminus Q$ and thus of trips to

    $F \smallsetminus N$. These trips intersect ∂N, which is compact, in a sequence $\{z_i\}$;

    let z be a cluster point of $\{z_i\}$. Since $\overline{N}$ is geodesically convex,

    there is a unique goedesic $[zx]$. The geodesic $[zx]$ cannot be timelike —

    if it were, we would have $z \in I^-(x)$ open, and thereby $z_i \in I^-(x)$

for some i, and then $X \gg z_i \Rightarrow x \in F$ which is false. Nor can $[zx]$ be spacelike — if it were, we could find a nbd. $U(x)$ lying outside $\mathcal{L}_z$ in N. $U(x) \ni y_i$, $\forall i >$ some i_o; also, $\exists y \in I^-(z)$ such that $\overline{I^+(y)} \cap U(x) = \phi$. But $I^+(y)$ contains z_i for some $i > i_o$. Hence the trip $y \rightarrow z_i \rightarrow y_i$ escapes from $\mathcal{L}_y$ in N, contradicting 2.7. Thus $[zx]$ is null. Lastly, $[zx]$ lies in B: since $\{z_i\} \subseteq F$, $z \in \overline{F}$. If $z \in F$, then there is $y \in F$ such that $y \ll z$; and $y \ll z \ll x \Rightarrow y \ll x \Rightarrow x \in F$, which is false. So $z \in \overline{F} \setminus F = B$. If $\omega \in [zx]$, then $I^+(\omega) \subseteq I^+(z)$; since $z < \omega$; thus $I^+(\omega) \subseteq F \Rightarrow \omega \in \overline{F}$. As above, $\omega \in F \Rightarrow x \in F$. So we have $[zx] \subseteq B$.

<u>Corollary</u>: Let S be a closed subset of M. Let $x \in \partial I^+[S] \setminus S$. Then there is γ, a null geodesic lying in $\partial I^+[S]$ with future endpoint x. Further, the maximal extension of γ into the past either intersects S, or γ is past endless on $\partial I^+[S]$.

<u>Proof</u>: Since $x \notin S$, which is closed, there is an open set Q containing x such that $Q \cap S = \phi$. If $y \in Q \cap I^+[S]$, there is a trip from y to S which must leave Q. Thus Q satisfies (a) of the proposition. Thus x is the future endpoint of a null geodesic in $\partial I^+[S]$. If γ is not past endless on $\partial I^+[S]$, it has an endpoint on $\partial I^+[S]$, say z. Then if $z \notin S$, we may repeat the above procedure to obtain another null geodesic on $\partial I^+ S$ with future endpoint z. Since γ is assumed to have been maximally extended, this new geodesic is not an extension of γ. Any point on it may be connected to x by a trip, contradicting the fact that $\partial I^+[S]$ is achronal.

§ 3: <u>STRONG CAUSALITY AND THE ALEXANDROFF TOPOLOGY</u>

3.1 The set of 4-dimensional manifolds admitting Lorentzian metrics is rather large and contains many "unphysical" models. To narrow this collection down some, it seems worth while to try to decide which, if any, of the local properties of M ought to be required to hold globally in a physically

reasonable model. One such property is generally agreed to be "causality".
There are no local violations of causality in any spacetime; if a<<a and γ
is a closed trip containing a, it is clear that γ must leave any simple
region containing a before intersecting a for the second time. Due to the
lack of any experimental evidence to the contrary, as well as for obvious
philosophical reasons, one often postulates the non-existence of closed
timelike curves. This is sufficient to rule out the possibility of compact
spacetimes:

3.2 Proposition: If M is compact, M has closed trips

Proof: We must show that there exists a point contained in its own future.
The collection $\mathcal{U} \equiv \{I^+(q) : q \in M\}$ is an open cover of M which, by
hypothesis, contains a finite subcover, say $\{I^+(q_1), \cdots , I^+(q_k)\}$.
Claim there is some q_i among these satisfying $q_i \in I^+(q_i)$. If not,
since $\{I^+(q_i) : 1 \leq i \leq k\}$ covers M, there is a q_{i_1} such that $q_1 \in I^+(q_{i_1})$,
$q_{i_1} \neq q_1$. Similarly, there is a $q_{i_2} \neq q_{i_1}$ such that $q_{i_1} \in I^+(q_{i_2})$.
Since $q_{i_2} << q_{i_1} << q_1$, $q_{i_2} \neq q_1$; otherwise $q_1 \in I^+(q_1)$. Proceeding in
this fashion, we obtain eventually $q_{i_{k-1}} << \cdots << q_{i_1} << q_1$. Now
$q_{i_{k-1}} \notin I^+(q_i)$ for any i, contradicting the fact that $\{I^+(q_i)\}$ is a
cover.

3.3 There are several types of causality violations which do not necessarily
entail the existence of closed timelike curves.
A spacetime M is said to be future-distinguishing if $a \neq b \Rightarrow I^+(a) \neq I^+(b)$
and past-distinguishing if distinct events have distinct pasts.

Remark: A spacetime may fail to be either future or past distinguishing and still
have no closed timelike curves. Take an Einstein cyllinder with future
null directions as indicated $\langle$rotation symmetry$\rangle$. $\left(\text{See fig. 9.}\right) \longrightarrow$

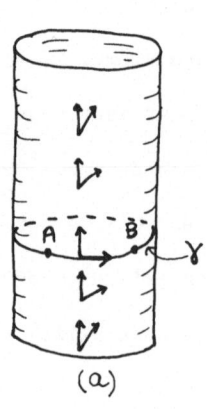 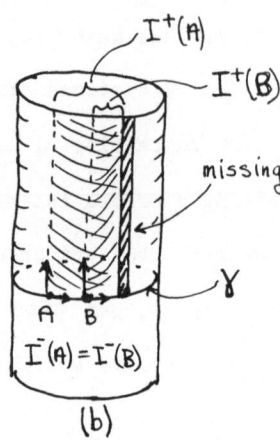

(a) (b)

fig. 9: The space in (a) has a closed causal curve γ through A and B, but no
closed timelike curves. $I^+(A) = I^+(B)$, $I^-(A) = I^-(B)$. Removing a point
of γ gives a spacetime with no closed causal curves which is neither
past-distinguishing nor future-distinguishing. Removal of the indicated
closed segment in (b) renders the space future-distinguishing, but it
still fails to be past-distinguishing.

3.4 A spacetime M is __strongly causal at a ε M__ <=> a has a neighborhood base
$\{\mathcal{U}_\alpha(a) : \alpha \in \Lambda\}$ with the property that __no__ $\mathcal{U}_\alpha$ is intersected by a trip in
a disconnected set. M is __strongly causal__ if it is strongly causal at each
point. See fig. 10.

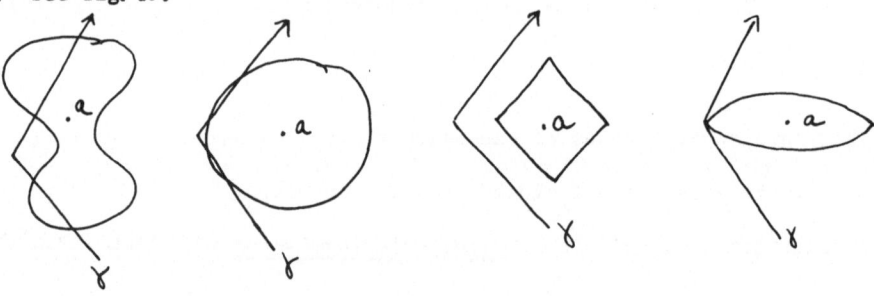

fig. 10: In M^2 ⟨and elsewhere⟩ one may construct various types of neighborhoods
at a point a. A neighborhood base at a constructed with sets similar to
either of the first two in the figure is not a suitable "causality base".
M^2 is strongly causal because there exists at each point a nbd. base
consisting of sets similar to either the third or fourth.

3.5 A strongly causal spacetime M is said to be <u>stably causal</u> ④, if it remains
strongly causal under slight variations in the metric. More precisely, there
exists a neighborhood of g in the bundle of Lorentz metrics such that (M, g')
is strongly causal for every g' in this nbd.

<u>Remark:</u>

    (a) It is necessary to require the existence of a nbd. <u>base</u> at a rather
    than just one such nbd.:

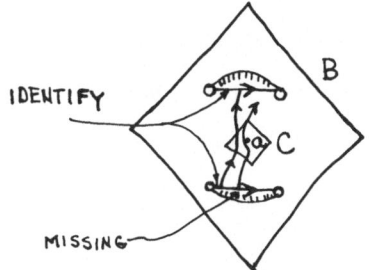

The causality violation is not necessarily
apparent in B, but there is no nbd. of a
contained in C which does not disconnect a
trip.

    (b) <u>Strong causality is not necessarily stable</u>; consider the following
    "baffle space" in which the lines l_1, l_2 and l_3 are missing:

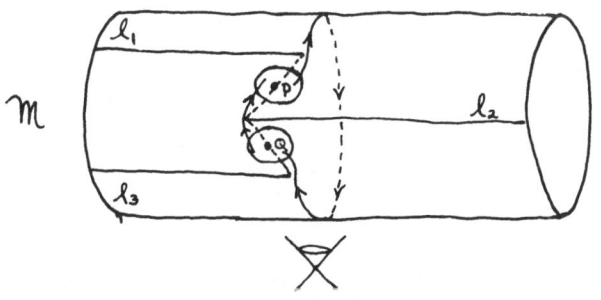

There are no closed causal
trips; further, M is
strongly causal. But it is
clear that a small change in
the metric which opens out
the light cones

will produce closed timelike curves at p ≠ q.

Since strong causality will play a major role in what follows, one is pleased
to have a precise statement of what it means for strong causality to fail:

3.6 <u>Proposition:</u> Let a ε M. Strong causality fails at a <=> there is a point
b > a, b ≠ a, such that for all $x \in I^+(a)$ and all y ε $I^-(b)$, y<<x.

Proof: Let N be a simple region containing a. Take any x>>a such that x and a are joined by a unique timelike geodesic in N. x is said to be in the $\underline{\text{local}}$ $\underline{\text{future}}$ of a, denoted $x \epsilon I_L^+(a)$. If y is in the local past of a, then $I_L^+(x) \cap I_L^-(y) = \{z : y << z << x$, where the trip lies in N$\}$ is clearly an open nbd. of a. If a trip intersects this nbd. in a disconnected set, it clearly must leave and reënter N in order to do so. If strong causality fails at a there must be a nested sequence of these nbds. at a, $\{U_i : i = 1,2,3\ldots\ldots \}$; $U_i \supset U_{i+1}$; $\bigcap_i \{U_i\} = \{a\}$; with the property that each U_i contains the past endpoint of a trip γ_i which leaves N and then comes back to enter U_i. Let $\{b_i\}$ be the sequence formed on the (compact) ∂N as the γ_i leave N. There is an accumulation point $b \epsilon \partial N$; we may select a convergent subsequence and renumber things so that $b_i \rightarrow b$. There exists a unique geodesic $[ab]$ lying in $\overline{N}$. $[ab]$ cannot be spacelike, so a<b. Now let $y \epsilon I^-(b)$, $x \epsilon I^+(a)$. There exists a nbd. $V(a) \subseteq I^-(x)$, and for some i_0, $i > i_0 \Rightarrow U_i \subseteq V(a)$. Also, there is a nbd. $W(b) \subseteq I^+(y)$ such that for some i_0', $i > i_0' \Rightarrow b_i \in W(b)$. Choose $i > i_0, i_0'$. Then $y_i << b_i$; b_i is on γ_i which enters $V(a)$ and thus $b_i << x$ So y<<x.

Conversely, suppose a and b exist satisfying the hypotheses. Separate a and b by disjoint nbds. $V(a)$ and $W(b)$. Let U(a) be any nbd. of a contained in $V(a)$. Choose $x^1 \epsilon I^-(a) \cap U(a)$ and $x \epsilon I^+(a) \cap U(a)$. Since $x^1 << a, + a < b$, $x^1 << b$. Thus there is a trip from x^1 to b which must leave $U(a)$. Choose y on this trip inside $W(b)$. They y<<x by hypothesis, and the trip $[x^1 yx]$ intersects $U(a)$ twice. Thus no nbd. of a contained in $V(a)$ can satisfy the strong causality condition and thus no nbd. base at a can satisfy it. So strong causality fails at a. Q.E.D.

Remark: If, given the conditions of the lemma, b is found to satisfy a<<b, then for any $y \epsilon I^-(b) \cap I^+(a)$, we have y<<y; i.e. there are closed trips in M.

3.7 The sets $\{I^+(a) \cap I^-(b) : a, b \epsilon M\}$ are open and clearly form a base for a topology on M, called the Alexandroff Topology. In general, the Alexandroff topology is coarser than the manifold topology. The condition for these two topologies to coincide is given in the following:

Theorem 1: A spacetime M is strongly causal <=> the Alexandroff topology is T_2 <=> the Alexandroff topology coincides with the manifold topology.

Proof: If strong causality fails at aϵM, then we find the point b of prop. 3.6. Clearly no two Alexandroff nbds. of a and b are disjoint; so the Alexandroff topology is not T_2. Hence the Alexandroff topology is strictly coarser than the manifold topology. Now if the Alexandroff topology is weaker than the manifold topology, there is an M - open set $U(y)$ such that any A-nbd. of y gets outside of U. Since M is regular we may assume this holds for $\overline{U}$ as well. Now let $V(y) \subset U$ be an arbitrary M - open nbd. of y. Let $b \epsilon I^+(y) \cap V(y)$, $a \epsilon I^-(y) \cap V(y)$. By hypothesis, $\{I^-(b) \cap I^+(a)\} \smallsetminus \overline{U} \neq \phi$. Choose z in this set. Then $z \notin V(y)$, but a<<z<<b, and $V(y)$ disconnects the trip [azb]. Since $V(y)$ is arbitrary, strong causality fails at y and the theorem is proved.

3.8 Examples of strong causality violations:

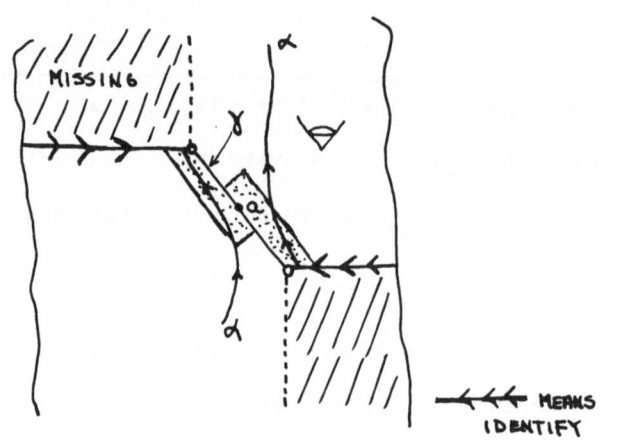

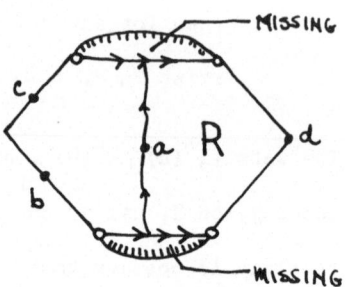

(a): Strong causality fails at a. There
      are no closed causal curves. γ
      is an endless null geodesic along
      which strong causality fails.
      The lightly shaded area is an A-nbd.
      of a.

(b): Strong causality is
      violated everywhere in R.
      There exist closed trips at
      a; there is a past endless
      geodesic at b along which
      strong causality is
      violated, a future endless
      one at c , and both at d .

3.9 <u>Proposition</u>: For each a ε M define $Q_a \equiv \{x \in \mathcal{M} :$ x lies on a closed trip
     through a}. Then Q_a is open, for every a ε M; and either $Q_a \cap Q_b = \phi$ or
     $Q_a = Q_b ,\ \forall\, a, b \in \mathcal{M}.$

<u>Proof</u>: Obvious.

One can be more specific about the structure of a region in which strong causality
is violated:

3.10 <u>Proposition</u>: If strong causality fails at a ε M, then one ⟨at least⟩ of
                    the following is true:

      (1) There are closed trips through a.

      (2) There is a future endless null geodesic through a along which strong
           causality fails; and there are closed trips near a.

      (3) Time reverse of (2).

(4)   There is an endless null geodesic through a along which strong
      causality fails.

Proof: Suppose strong causality fails at a.   Then we find the point b of
prop. 3.6.   Suppose first that $a << b$.   Then by the remark following
3.6 there are closed trips arbitrarily close to a.   Suppose there are
no closed trips through a.   Choose $y \in I^-(b) \cap I^+(a)$ and set $Q = Q_y$
Clearly $a \in \partial Q$.   Let $F = I^+[Q]$, $P = I^-[Q]$.   Note that $Q = F \cap P$
and that $\partial Q \subset \partial F \cup \partial P$.   Now $a \in \partial Q$ and there is a nbd. $U(a)$
such that $U \cap Q \nsubseteq Q$.   Let $z \in Q \setminus U$.   Then any point in $Q \cap U$
has a trip to z .   If $a \in \partial F$,   then a satisfies the conditions
in prop 2.22 and a is the future end point of a null geodesic lying
in ∂F, say γ.   Let $c < a$ lie on γ .   Then it is immediate by
prop 3.6 that strong causality fails at c as long as $c \in \partial F$ If γ
is not past endless on ∂F,   it has a past end point in $\bar{Q}$ by the
corollary to 2.22; but this end point cannot be in Q itself, since
we would then be able to construct closed trips through a.   So it
must lie on ∂Q .   Since we may repeat this construction, starting
at this point, it follows that we may continue γ and that it still
remains in ∂Q.   It is past-endless on ∂Q.   If $a \in \partial P$ the
time-reverse of 2.22 produces (2).   The remaining possibility is that
$a < b$ and $a +\!\!+ b$, $c < a$ with $c +\!\!+ a$. $\langle$from the time-reverse
of 2.22$\rangle$.   It is clear that strong causality must also fail at every
point of the endless null geodesic containing the segment $[cab]$ and
so (4) holds.

§ 4: FUTURE-TRAPPED SETS; DOMAINS OF DEPENDENCE; CAUCHY HORIZONS.

4.1 Define the set $E^+[S] = J^+[S] \setminus I^+[S]$.   Clearly, $E^+[S] \subseteq \partial I^+[S]$.   If S is
closed, $E^+[S]$ is the portion of $\partial I^+[S]$ which may be connected to S by
null geodesics.

4.2. <u>Examples</u>:

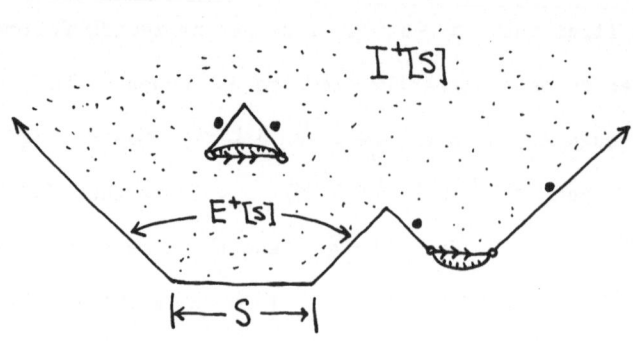

(a) Mutilated M^2; $E^+[s]$ is indicated in the figure. Segments marked with • lie on $\partial I^+[s]$ but not on $E^+[s]$.

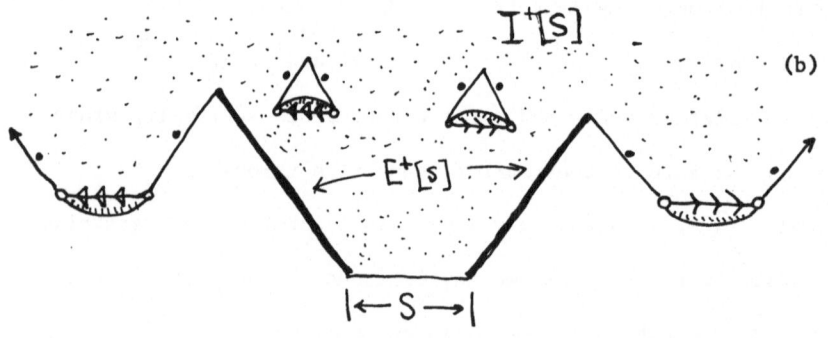

(b) This is obtained by "doubling" figure (a). Note that $E^+[s]$ is compact. S is future trapped.

4.3. A non-empty, closed achronal set S is said to be <u>future-trapped</u> if $E^+[s]$ is compact. Past-trapped sets are defined analogously.

4.4. Let $S \subseteq M$. Define $D^+(s) \equiv \{ x \in M :$ every past-endless trip through x meets $s \}$; $D^-(s) \equiv \{ x \in M :$ every future-endless trip through x meets $s \}$; $D(s) \equiv D^+(s) \cup D^-(s)$. $D(s)$ is called the <u>domain of dependence</u> of S while $D^+(s)$ and $D^-(s)$ are called the future and past domains of dependence.

4.5 Examples:

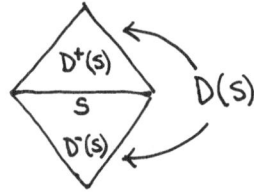

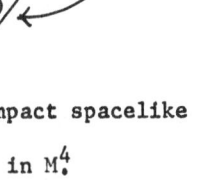

missing

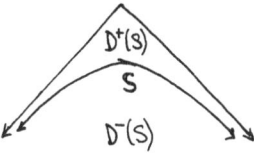

(a) S is a compact spacelike

3-surface in M^4.

[2 spatial dimensions have been suppressed.]

(b) S is the same as in

(a), only a pt. has

been removed from

M^4.

(c) S is half of

the unit

pseudosphere

in M^4.

4.6 The Cauchy horizon of S, denoted $H(S)$, is defined by $H(S) = H^+(S) \cup H^-(S)$,

where $H^+(S) \equiv \{x : x \in D^+(S), I^+(x) \cap D^+(S) = \emptyset\}$, $H^-(S) = \{x \in D^-(S) : I^-(x) \cap D^-(x) = \emptyset\}$.

4.7 Example: $H(S)$ will, in general, include points of S:

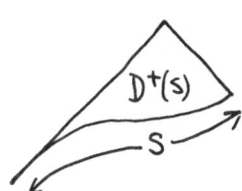

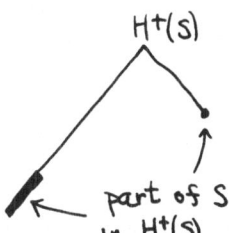

$H^+(S)$

part of S in $H^+(S)$

Remark: S will be assumed achronal and closed for the remainder of section 4.

4.8 Proposition: Let $S \subset M$ be achronal and closed. Then:

(1) $H^+(S)$ is achronal and closed. $\langle H^+(S)$ may be empty.$\rangle$

(2) $D^+(S)$ is closed.

(3) $x \in D^+(S) \Rightarrow \{I^-(x) \cap I^+[S]\} \subset D^+(S)$.

(4) $\partial D^+(S) = H^+(S) \cup S$.

(5) $I^+[H^+(S)] = I^+[S] \setminus D^+(S)$.

Proof: exercise.

4.9 Let $S \subset \mathcal{M}$ be achronal and closed. The <u>edge of S</u> is defined by edge $S \equiv$ $\{x \in S: \forall y, z \mid z \ll x \ll y$, there is a trip $[zy]$ not meeting $S.\}$

$x \in$ edge (S). edge (S) is the set of points at which S fails to be a topological 3-manifold.

For example, if S is a spacelike line in M^4, edge (S) = S.

4.10 <u>Proposition</u>: (a) $x \in D^+(S) \Rightarrow I^-(x) \cap edge\ (S) = \emptyset$

(b) edge (S) = edge $\left(H^+(S)\right)$.

4.11 <u>Proposition</u>: Let $x \in H^+(S) \setminus edge(S)$. Then there is a null geodesic on $H^+(S)$ with future endpoint x.

<u>Proof</u>: Set $F = I^+[H^+(S)]$. $x \in \partial F$. Since $x \notin$ edge (S), we can find a nbd. $Q(x)$ satisfying the conditions of 2.22. Done.

4.12 <u>Proposition</u>: $x \in D^+(S) \setminus H^+(S) \implies$ every past endless causal trip through x intersects $S \setminus H^+(S)$.

<u>Proof</u>: Let γ be a past endless null geodesic with future endpoint x. If $x \in S$, then we are finished. If not, by 4.8 (3), $x \in int\ D^+(S)$ and thus $I^+[x] \cap D^+(S) \neq \emptyset$. If $y \in I^+(x) \cap D^+(S)$, then γ is in the past of y and must intersect S. γ cannot intersect $H^+(S)$, for this would contradict the definition of $H^+(S)$.

4.13 <u>Proposition</u>: $\partial D(S) = H(S)$

4.14 <u>Proposition</u>: $x \in D^+(S) \Rightarrow \{I^-(x) \cap J^+[S]\} \subseteq D^+(S)$.

4.15 <u>Proposition</u>: $x \in int\ D^+(S) \implies$ strong causality holds at x.

<u>Proof</u>: By 4.12, every past-endless null geodesic through x intersects S. If strong causality fails at x, it thus fails at a point $s \in S \setminus H^+(S)$.

It follows from 3.6 that S is not achronal, which is a contradiction.

4.16 **Proposition:** $x \in \text{int } D^+(S) \implies J^-(x) \cap J^+[S]$ is compact.

Proof: Assume $K = J^-(x) \cap J^+[S]$ is not compact. Then there is an open cover $\{U_i : i = 1, 2, 3 \cdots\}$ which is countable, locally finite, has no finite subcover and whose elements are simple regions.* Choose $a_i \in U_i$ for each i. The sequence $\{a_i\}$ has no cluster points in K since $\{U_i\}$ is locally finite. Construct trips $\{\gamma_i\}$ where γ_i goes from x to a_i to S. Let U_{i_1} be an element of the cover containing x. ∂U_{i_1} is compact and the intersections $\{\gamma_i \cap \partial U_{i_1}\}$ have a cluster point $x_1 \in \partial U_{i_1}$. Choose $z_0 \in I^+(x) \cap U_{i_1}$. Then

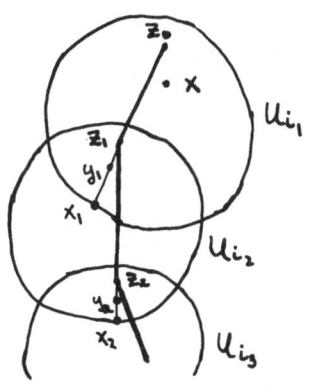

$z_0 \gg x_1$. Construct the timelike geodesic $[z_0 x_1]$. There is $i_2 \neq i_1$ such that $x_1 \in U_{i_2}$. Choose y_1 on $[z_0 x_1]$ in $U_{i_1} \cap U_{i_2}$. Infinitely many of the a_i are in $I^-(y_1)$. Thus the past exit points of the trips $\{\gamma_i\}$ accumulate at a point $x_2 \in \partial U_{i_2}$. Choose z_1 on $[z_0 x_1]$ such that $z_1 \in U_{i_2}$; $z_1 \gg y_1$. Then choose y_2 on $[z_1 x_2]$ in $U_{i_2} \cap U_{i_3}$ where $U_{i_3} \ni x_2$, and proceed to obtain x_3,

and then z_2. And so on. The process is endless because:

1) The trip $[z_0 z_1 z_2 \cdots]$ cannot meet S — otherwise we would have infinitely many of the a_i in a compact set.

---

*M may be topologically embedded in a finite dimensional Euclidean space, so M is Lindelöf. If K is non-compact, there is an open cover with no finite subcover; by paracompactness, there is a locally finite open refinement of this; since the space is Lindelöf, there is a countable subcover of this. Since the simple regions form a basis, this final cover may be taken to consist of simple regions.

2) The trip is past endless in K— if b were a past endpoint b would lie in U_i for some i, so this U_i would contain infinitely many of the $z_i's$, contradicting the fact that $\{U_i\}$ is locally finite.

So, we have a point $z_0 \in D^+(S)$, and a past-directed trip from z_0 which does not intersect S when maximally extended. This contradicts the definition of $D^+(S)$. Thus $J^-(x) \cap J^+[S]$ is compact, as asserted.

§ 5: THE SPACE OF CAUSAL CURVES

Given p and q in M such that $p \ll q$, it is important to know if there is a timelike geodesic connecting them, and if so, whether or not this geodesic is maximal. The existence of such a geodesic is not implied by geodesic completeness as it is in the Riemannian case; a well-known counter example is the universal covering space of two-dimensional anti - de Sitter space: This space is geodesically complete, but the future of any given point contains many points which cannot be joined to it by a geodesic. The purpose of this section is to establish a sufficient condition for the existence of causal curves of maximal length.

5.1 In a spacetime M, let $G \subset M$ be defined by $G = \{x: M$ is strongly causal at $x\}$.

5.2 Proposition: G is open

> Proof: Suppose $\exists x \in M$, and $\{x_i\} \subset M \setminus G$, such that $x_i \longrightarrow x$. Take a simple region containing x. For each x_i, find the point b_i predicted by 2.22. Take the intersection of the geodesic $[x_i b_i]$ with the (compact) boundary of the region. These intersections have a cluster point y. Take any two disjoint neighborhoods of x and y. Using 2.22 it is easy to construct a trip leaving this neighborhood of x, entering the nbd. of y and then reëntering the nbd. of x, showing that strong causality fails at x. Thus $x \notin G$.

5.3 For the remainder of this section, it is assumed that A and B are closed achronal sets contained in K, a compact subset of G. We denote by $\mathcal{C}_K(A,B)$ the set of causal curves from A to B lying in K, and by $\mathcal{T}_K(A,B)$ the set of causal trips from A to B lying in K. We will write $\mathcal{C}(A,B)$ for $\mathcal{C}_m(A,B)$.

5.4 M, as a topological space, is metrizable. Choose an arbitrary metric d compatible with the manifold topology. Then in G, d is compatible with the Alexandroff topology.

Recall that for a set E and a point p, $d(p,E) \equiv \inf\{d(p,y): y \in E\}$. For $\gamma \in \mathcal{C}_K(A,B)$, define $V_\epsilon(\gamma) \equiv \{y \in G: d(y,\gamma) < \epsilon\}$. For $\gamma_1, \gamma_2 \in \mathcal{C}_K(A,B)$, define $\rho(\gamma_1,\gamma_2) \equiv \inf\{\epsilon: V_\epsilon(\gamma_1) \supset \gamma_2 \text{ and } V_\epsilon(\gamma_2) \supset \gamma_1\}$. ρ is the restriction to $\mathcal{C}_K(A,B)$ of the <u>Hausdorff</u> <u>metric</u>. $\langle \rho$ is a pseudo-metric on the power set of M and a metric on the closed subsets of M. The elements of $\mathcal{C}_K(A,B)$ are compact and thus closed.$\rangle$

5.5 <u>Proposition:</u> $\mathcal{T}(A,B)$ is dense in $\mathcal{C}(A,B)$.

<u>Proof:</u> Let $\gamma \in \mathcal{C}(A,B)$. Choose $\epsilon > 0$. Cover γ by finitely many $\frac{\epsilon}{2}$-balls. The union of these is a nbd of γ, and by the defn. of causal curve, there is a causal trip joining the endpoints of γ and lying in this neighborhood, say μ. Clearly $\mu \in V_\epsilon(\gamma)$ and since the distance from any pt. of γ to μ is less than ϵ, $\gamma \in V_\epsilon(\mu)$. Thus $\rho(\mu,\gamma) < \epsilon$, and $\mathcal{T}$ is dense in $\mathcal{C}$. This is the precise formulation of the fact that the causal curves form the complete set of pointwise limits of the causal trips. Note that the assumption of the achronality of A and B was not used here.

5.6 Recall that a metric space is compact iff it is complete and totally bounded — $\langle \underline{X}$ is totally bounded if the Cover $\{S_\epsilon(x): x \in \underline{X}\}$ has a finite subcover for any $\epsilon > 0.\rangle$

5.7 <u>Proposition</u>: If K is compact and contained in G, $\ell_{e_K}(A,B)$ is totally bounded.

<u>Proof</u>: Let $\varepsilon > 0$ be given. K is compact, hence totally bounded and may therefore be covered by finitely many Alexandroff nbds. of d-diameter less than $\frac{\varepsilon}{2}$, say $\{A_1, A_2, -- , A_k\}$. Form the finite collection of open sets $\{B_i\}$ where each B_i is a chain of A_j's, $A_{j_i} \cup \ldots \cup A_{j_n}$, satisfying $A \cap A_{j_i} \neq \phi$, $A_{j_n} \cap B \neq \phi$, $A_{j_k} \cap A_{j_{k+1}} \neq \phi$, and $I^+[A_i] \cap A_{i+2} \neq \phi$.

Set $B_i' \equiv \{\gamma \in \ell_{e_K}(A,B) : \gamma < B_i\}$. By construction, we have $B_i' \neq \phi$, each i; $\cup \{B_i'\} = \ell_{e_K}(A,B)$; $\gamma \in B_i' \Rightarrow S_\varepsilon(\gamma) \supset B_i'$. Choose $\gamma_i \in B_i'$ for each i. Then the collection $\{S_\varepsilon(\gamma_i)\}$ so determined is a finite cover of $\ell_{e_K}(A,B)$.

5.8 <u>Proposition</u>: $\ell_{e_K}(A,B)$ is complete.

<u>Proof</u>: Let $\{\gamma_i\} \subseteq \ell_{e_K}(A,B)$ be a Cauchy sequence. Cover K by finitely many Alexandroff nbds. each of which is contained in a simple region whose closure lies in G. Let $\{x_i\}$ be the sequence on A formed by the intersections of the γ_i. $\{\gamma_i\}$ is Cauchy, and A is compact $\Rightarrow x_i \rightarrow x_0 \in A$. Let A_1 be an Alexandroff nbd. in the cover which contains x_0. Introduce Minkowskian coordinates in $\overline{A}_1$ centered at x_0. Since $x_0 \in \mathring{A}_1$ there is some interval $[0, t_1)$ such that for each t satisfying $0 \leq t < t_1$, the compact 3-surface $\{t = \text{constant}\} \cap \overline{A}_1$ is intersected by infinitely many γ_i in points $x_i(t)$ in A_1. For each such t, $\{x_i(t)\}$ is a Cauchy sequence in the compact 3 - surface and thus converges to a limit which we denote $x_0(t)$. It is clear that $0 < t' < t'' < t_1 \Rightarrow x_0(t') < x_0(t'')$, and that $x_0(t)$ is continuous - for example, if $x_0(t)$ fails to be continuous from below at t', then there is an Alexandroff nbd of $x_0(t')$ containing no other $x_0(t)$ for $0 < t < t'$. This is impossible

since infinitely many of the γ_i's must pass through this nbd .
Now either $X_0(t)$ intersects B inside of A_1, in which case we are
finished, or there is a first accumulation point on ∂A_1, say $X_0(t_1)$
$\langle$the Minkowskian coordinates are valid on ∂A_1, since $\overline{A_1}$ is
contained in a simple region$\rangle$. Let A_2 be an Alexandroff nbd. in
the cover which contains $X_0(t_1)$. Repeat the process, taking the
Minkowskian coordinates in A_2 to have the value $(t_1,0,0,0)$ at $X_0(t_1)$
The process is finite since the curve $X_0(t)$ so constructed is a
causal curve and cannot re-enter any Alexandroff nbd. it has left.
$X_0(t)$ must strike B, since B is closed. Done.

Theorem 2: $\mathcal{C}_{eK}(A,B)$ is a compact metric space.

Proof: Q.E.D.

Corollary 1: Let S be closed, achronal, and strongly causal. Let $x \in \text{int } D^+(S)$.
Then $\mathcal{C}_e(x,S)$ is compact.

Proof: $\mathcal{C}_e(x,S) \subset K \equiv J^-(x) \cap J^+[S]$, which is compact and contained in G.

Corollary 2: Let S be closed, achronal, and strongly causal. Let $a,b \in \text{int } D(S)$.
Then $\mathcal{C}_e(a,b)$ is compact.

Proof: Obvious. Note that the corollary is false if we allow $a \in H(S)$:

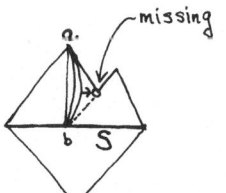

The sequence indicated is a Cauchy sequence in $\mathcal{C}_e(a,b)$,
but it is not convergent, since there is a point
missing. This cannot happen in the interior of $D(S)$.

5.9 A spacetime M is said to be <u>globally</u> <u>hyperbolic</u> iff M is strongly causal and
$\mathcal{C}_e(a,b)$ is compact, for all $a,b \in \mathcal{M}$.

5.10 A <u>Cauchy hypersurface</u> for $\mathcal{M}$ is a closed achronal set S for which $D(S) = \mathcal{M}$.

5.11 <u>Proposition</u>: $\mathcal{M}$ has a Cauchy hypersurface $\Longrightarrow$ $\mathcal{M}$ is globally hyperbolic.

    <u>Proof</u>: Corollary 2

5.12 Examples:

    a) In $\mathcal{M}^4$, the surface $t=6$ is a $C.H.S.$

    b) The pseudosphere $t^2-x^2-y^2-z^2=1, \; t<0$ is not, although the surface is closed, achronal, and non compact – See example 4.5 (c)

    c) $\mathcal{M}^4 \smallsetminus \{pt.\}$ does not have a $C.H.S.$

<u>Remark</u>: The requirement that a spacetime possess a Cauchy hypersurface is very strong, as example c) indicates. It has been shown $([2])$ that global hyperbolicity is equivalent to the existence of a $C.H.S.$; also $([2])$ that if M has a $C.H.S.$ S, M is homeomorphic to $\mathbb{R} \times S$. The homeomorphism may be constructed in such a way that $f(t_0, S)$ is a $C.H.S.$ for every $t_0 \in \mathbb{R}$. It is clear that there are spacetimes homeomorphic to $\mathbb{R} \times S$ (S a 3 – dimensional manifold) which do not have Cauchy hypersurfaces.

6: <u>THE SIGNIFICANCE OF CONJUGATE POINTS; SINGULARITY THEOREMS.</u>

We have already seen in § 1 that $\int ds$ is maximized locally along timelike geodesics. Another way to see this is to introduce what is known as a <u>synchronous coordinate system</u>, generated by a spacelike hypersurface and a hypersurface orthogonal family of timelike geodesics. At a point $m \in \mathcal{M}$ take Minkowskian coordinates in T_m and transform them according to: $(\longrightarrow)$

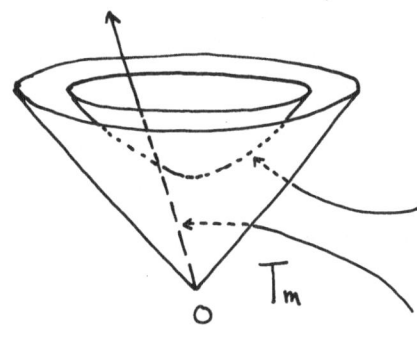

$$x^1 = X = \frac{x}{t}$$

$$x^2 = Y = \frac{y}{t}$$

$$x^3 = Z = \frac{z}{t}$$

$$T = \sqrt{t^2 - x^2 - y^2 - z^2}$$

T_m

O

A spacelike hypersurface corresponding to $T =$ const.

A timelike geodesic corresponding to $X = X_o, Y = Y_o, Z = Z_o$; T varying.

The X, Y, Z, T coordinates are valid in the interior of the future (past) null cone in T_m and exp_m maps them into a synchronous coordinate system in the immediate future of m. T is an affine parameter along the radial future-directed timelike geodesics at m and measures proper time. The images under exp_m are the hypersurfaces $S_T(m)$ of Lemma 4. The synchronous coordinates are valid in the region of $I^+(m)$ in which exp_m is a diffeomorphism.

Since T is an affine parameter with $g\left(\frac{\partial}{\partial T}, \frac{\partial}{\partial T}\right) = 1$, and since the radial geodesics are the orthogonal trajectories of the $T =$ constant surfaces, $\left(g\left(\frac{\partial}{\partial T}, \frac{\partial}{\partial x^j}\right) = 0\right)$, the metric takes the form $ds^2 = dT^2 - \gamma_{ij} dx^i dx^j$ where at each point $(\!(\gamma_{ij})\!)$ is a positive definite symmetric matrix. If we consider a smooth causal curve γ from a point of $T = a$ to a point of $T = b$, we have

$$L(\gamma) = \int_\gamma ds = \int_\gamma \sqrt{dT^2 - \gamma_{ij} dx^i dx^j} \leq \int_{T=a}^{T=b} dT = b - a \text{, with}$$

equality holding only for $x^1 = x^2 = x^3 =$ constant, i.e only when γ is a geodesic.

6.1 Let γ be a causal trip. The <u>length of γ</u>, $L(\gamma)$, is defined to be the sum of the lengths of the segments constituting γ.

6.2 <u>Proposition</u>: Let A and B be achronal topological manifolds in a strongly causal spacetime $\mathcal{M}$. Then the length function is upper semi-continuous on $\gamma(A, B)$.

<u>Proof</u>: Let $\gamma_k \to \gamma_0$ in $\Upsilon(AB)$. It must be shown that $\limsup \{L(\gamma_k)\} \leq L(\gamma_0)$. Choose a subsequence of $\{\gamma_k\}$ and relabel so that $L(\gamma_k) \longrightarrow a = \limsup \{L(\gamma_k)\}$. It is sufficient to prove the proposition in the case that γ_0 is a geodesic and $\Upsilon(A,B) \subset N$, a simple region.

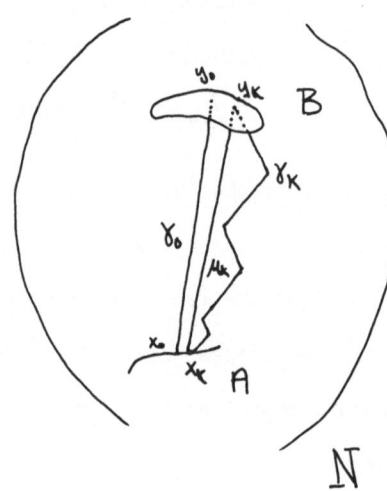

Let x_k, y_k be the intersections of γ_k with A and B respectively; and let μ_k be the unique causal geodesic $[x_k y_k]$ lying in N. Since $\gamma_k \to \gamma_0$, we have $x_k \to x_0$ and $y_k \to y_0$.

Also, since L is a <u>continuous</u> function on geodesics joining two continuous surfaces, we have $L(\mu_k) \longrightarrow L(\gamma_0)$. By the local maximality of geodesics, for every k we have $L(\gamma_k) \leq L(\mu_k)$. Thus $\lim \{L(\gamma_k)\} = a \leq \lim \{L(\mu_k)\} = L(\gamma_0)$, and so the superior limit of the original sequence is less than or equal to $L(\gamma_0)$.

<u>Remark</u>: L is <u>not</u> continuous on $\Upsilon(A,B)$; a sequence of null curves may converge to a timelike curve:

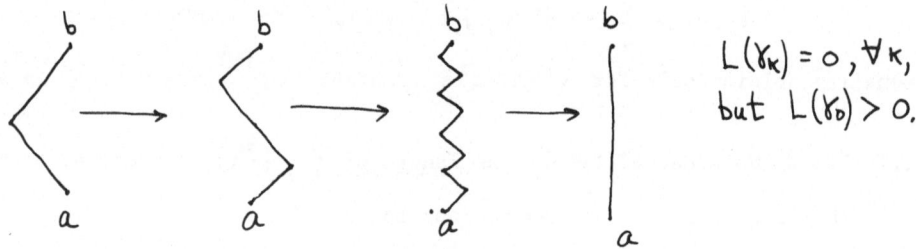

$L(\gamma_k) = 0, \forall k,$ but $L(\gamma_0) > 0.$

6.3 Since $\Upsilon(A,B)$ is dense in $\mathcal{C}_\ell(A,B)$, L has an upper semi-continuous extension to $\mathcal{C}_\ell(A,B)$, defined by setting $(\longmapsto)$

$$L(\gamma) = \lim_{\substack{\mu \to \gamma \\ \mu \in \gamma(A,B)}} \sup \left\{ \sum L(\mu) \right\}$$

The restriction of L to $\mathcal{C}_K(A,B)$, where K is any compact set, remains upper semi-continuous. So we have immediately:

<u>Theorem 3</u>: There is a curve $\gamma \in \mathcal{C}_K(A,B)$ with maximal length.

<u>Proof</u>: An upper semi-continuous function on a compact set attains its maximum. Note that γ needn't be unique and needn't be a geodesic:

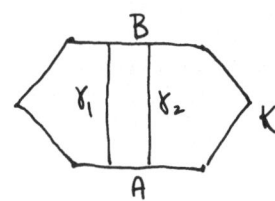  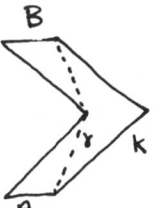

If A and B are parallel in M^4 then γ_1 and γ_2 are maximal.

Here there is only one element in $\mathcal{C}_K(A,B)$ which is necessarily maximal. It is not a geodesic.

Here the maximal element is a trip.

6.4 It is clear that any segment of such a maximal curve which lies in the interior of K must be a geodesic segment. In a globally hyperbolic spacetime, $a \ll b$ $\Rightarrow$ there is a timelike geodesic from a to b which has maximal length.

6.5 Recall the definition of Jacobi fields. Two points m and n on a geodesic γ are said to be <u>conjugate along</u> γ iff. there is a non-trivial $\langle$not identically zero$\rangle$ Jacobi field on γ vanishing at m and n . Equivalently, if X is the element of T_m such that $\exp_m(tX) = \gamma$, and $\exp_m X = n$, n is conjugate to m along γ iff $\exp_m$ is singular at X $\langle$ the differential map does not have full rank at $X\rangle$.

6.6 One may also define conjugate points to a surface: if S is a smooth spacelike surface, construct the congruence of future-directed timelike geodesics normal to S. The affine parameter along these geodesics is uniquely specified by requiring $t=0$ on S and $g(T,T)=1$, where $T=\frac{\partial}{\partial t}$ Choosing arbitrary coordinates (x^1, x^2, x^3) on S, we may introduce coordinates in $I^+[S]$ by putting $\gamma(t) = (t, x^1, x^2, x^3)$, where $\gamma(0) = (x^1, x^2, x^3)$ The geodesics $\{\gamma\}$ are the orthogonal trajectories of the spacelike surfaces $t =$ constant.

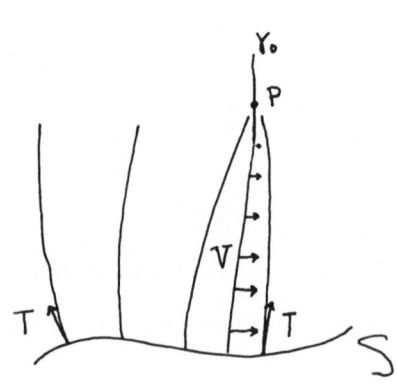

A point p is said to be conjugate to S along γ_0 iff there is a non-trivial Jacobi field V along γ_0 vanishing at p. V is assumed to be the variation vector field of a one-parameter variation through $\{\gamma\}$.

The significance of conjugate points lies in the fact $\langle$ to be proven $\rangle$ that beyond a conjugate point, a causal geodesic is no longer the maximal curve joining its end points $\left(\begin{array}{l}\text{or, in the case of a surface, no}\\ \text{longer the max. curve joining the surface}\\ \text{to its future endpoint.}\end{array}\right)$

6.7 Proposition: Let $V(t)$ be a non-trivial Jacobi field along $\gamma(t)$ such that $V(0) = 0$, $\frac{DV}{dt}\big|_{t=0} = A \neq 0$. Let A' be the associated constant vector field in $T_{\gamma(0)}$. Then

$$\exp_{p*}[tA']\big|_{\exp_p^{-1}[\gamma(t)]} = V(t).$$

Proof: The map $(t, w) \xrightarrow{f} \exp_{\gamma(0)}\left[t(T+wA)\right]$ is a 1 - parameter variation of geodesics and clearly

$$\frac{\partial}{\partial w}(t,0) \xrightarrow{df} \exp_{p*}[tA']\big|_{\exp_p^{-1}[\gamma(t)]}.$$ So $\exp_{p*}[tA']$ is a Jacobi field. Moreover $\exp_{p*}[tA'] = 0$ at the origin and

$$\frac{D}{dt}\left[\text{exp}_{P*}[tA']\right]\Big|_{t=0} = \frac{D}{dt}\left\{t \cdot \text{exp}_{P*}A'\right\}\Big|_{t=0} =$$

$$\text{exp}_{P*}A'\Big|_{t=0} + t\frac{D}{dt}\left[\text{exp}_{P*}A'\right]\Big|_{t=0} = A = \frac{DV}{dt}\Big|_{t=0}. \quad \text{And by}$$

the uniqueness of Jacobi fields, the proposition is established.

Theorem 4:

(a) Let S be a smooth spacelike hypersurface and $\{\gamma\}$ the orthogonal
congruence described above. If p is conjugate to S along γ_0
and if $r \gg p$ on γ_0, then there is a trip from S to r
of length greater than that measured on γ_0.

(b) If p is conjugate to q on $\gamma(t)$, a timelike geodesic, with $p \gg q$,
and if $r \gg p$ on γ, then there is a trip from q to r of length
greater than that measured on γ.

(c) If p is conjugate to q on a null geodesic γ with $p > q$, then
for $r > p$ on γ, there is a trip from q to r. - i.e. $r \in I^+(q)$.

Proof: We will prove (a). The proof of (b) is almost identical. The proof
of (c) uses the same general ideas, but with considerably greater
finesse. For complete details, the reader should consult [11].
To prove (a) first relabel the affine parameter such that $\gamma_0(0) = p$.
Let N be a simple region about p and choose $r \gg p$ on γ_0 with $r \in N$.
      Take a synchronous coordinate system for the past of r with the
hypersurfaces $S_{\bar{t}}(r)$ labelled so that $\bar{t} = t$ along γ_0. Note that
the function $t - \bar{t}$ is well-defined and smooth to the past of p.
The object of the proof is to produce a point at which $t - \bar{t} > 0$.
Let $V(t)$ be a non-trivial Jacobi field along γ_0 with $V(0) = 0$.
Then $\frac{DV}{dt}\Big|_{t=0} \neq 0$.   With the notation of prop. 6.7, let $W(t)$
denote the field $\text{exp}_{P*}A'$ along $\gamma_0(t)$. Then $V(t) = tW(t)$; $g(V,T) = 0$
$\Rightarrow g(W,T) = 0$, so W is spacelike, and by construction, $W(t) \neq 0$
between S and p. For $t < 0$, $W = t^{-1}V$ and $\nabla_W T = \nabla_{t^{-1}V} T$
$= t^{-1}\nabla_V T = t^{-1}\nabla_T V = t^{-1}\nabla_T(tW)$
$= t^{-1}\left\{W + t\frac{DW}{dt}\right\} = t^{-1}W + \frac{DW}{dt}.$ Taking the

scalar product with W, we have

$$g(\nabla_W T, W)\big|_t = t^{-1} g(W, W)\big|_t + g\left(\frac{DW}{dt}, W\right)\big|_t \, .$$

Since W is spacelike, $\lim\limits_{t \to 0^-} g(\nabla_W T, W) = +\infty$. $\Big[$This is acceptable since T is not defined at $t=0$, because the coordinate system is singular there.$\Big]$ On the other hand, letting $\hat{T} = \frac{\partial}{\partial t}$ we have $T = \hat{T}$ on γ_0 for $t < 0$; since $\hat{T}$ is well defined at $t=0$, $g(\nabla_W \hat{T}, W)$ is well behaved along γ_0 in a nbd. of p. So there is a point $m \ll p$ on γ_0 at which $g(\nabla_W T, W) - g(\nabla_W \hat{T}, W)$

$$= g(\nabla_W [T - \hat{T}], W) > 0 \, .$$

Construct $\mu(s)$, the unique geodesic through m satisfying $\mu(0) = m$, $\frac{d\mu}{dt}\big|_{t=0} = W(m)$, and define W on μ by $W(s_0) = \frac{d\mu}{ds}\big|_{s_0}$. Then W, T and $\hat{T}$ are all defined on $\mu(s)$, and by continuity there is an $\varepsilon > 0$ such that $s \in (-\varepsilon, \varepsilon) \implies$

$$g(\nabla_W [T - \hat{T}], W) > 0 \, . \qquad \text{Now on } \mu \text{ we have } \nabla_W W = 0 \, , \text{ so that}$$

$$g(\nabla_W [T - \hat{T}], W) = \nabla_W \big[g(T - \hat{T}, W) \big] \, . \qquad \text{Observing that}$$

$$g(T, W) = \langle W, dt \rangle = \nabla_W(t) \, , \qquad \text{and similarly for } \hat{T},$$

we have $\nabla_W \big[g(T - \hat{T}, W) \big] = \nabla_W^2 (t - \bar{t}) > 0$ for $s \in (-\varepsilon, \varepsilon)$. Write $t - \bar{t} = (t - \bar{t})(s)$ on μ. Expanding in a Taylor series about $s = 0$, we have $(t - \bar{t})(s) = (t - \bar{t})(0) +$

$$\left[\frac{d}{ds}\Big|_{s=0} (t - \bar{t}) \right] s + \left[\frac{d^2}{ds^2}\Big|_{s=0} (t - \bar{t}) \right] \frac{s^2}{2!} + 0(s^3) \, . \text{ The first term is 0.}$$

If the first derivative $\neq 0$ at $s = 0$, then $(t - \bar{t})(s)$ is monotone in a nbd. of 0, so there is a point for which $(t - \bar{t}) > 0$. If the first derivative is 0 at $s = 0$, then $\frac{d^2}{ds^2}(t - \bar{t}) \frac{s^2}{2!} = $

$$\big[\nabla_W^2 (t - \bar{t}) \big] s^2/2! > 0 \implies (t - \bar{t}) \qquad \text{has a local minimum there.}$$

Thus in either case, there is a point n to the past of p for which $t - \bar{t} > 0$. Let the length of γ_0 from s to r equal c. At n, we have $t = \bar{t} + \delta$. The geodesic $\gamma(t)$ from s to n has length t; the geodesic γ' from r to n has length $c - \bar{t} = $

$= c - t + \delta$. Thus the length of the trip $\gamma \cup \gamma'$ has length $c - \bar{t} + t = c + \delta > c$, and the statement is proved.

Corollary 1: Let F be a future set, γ a null geodesic on ∂F. Then γ contains no proper segment which has a pair of conjugate points.

  Proof: If $x < y < z$ along γ with x and y conjugate, then

$$z \in I^+(x) \implies z \in F \implies z \notin \partial F \qquad \text{which is false.}$$

Corollary 2: Let $\mathcal{M}$ be a spacetime satisfying:

  (1) There are no closed trips

  (2) <u>Every</u> endless null geodesic in $\mathcal{M}$ contains a pair of conjugate points.

  Then $\mathcal{M}$ is strongly causal.

  Proof: If not, and there are no closed trips, strong causality fails on an endless null geodesic γ. By hypothesis, there are conjugate points A and B on γ with $A < B$. Choose $a < A$, $b > B$. Then strong causality fails at a and b, and by the theorem, $I^+(a) \cap I^-(b) \neq \emptyset$ and there are closed trips $\langle$ see the remark following 3.6$\rangle$, which is false.

6.8 **Proposition:** Let S be future trapped and suppose strong causality holds in $\overline{I^+[S]}$. Then there is a future endless trip γ such that $\gamma \subset \text{int } D^+(E^+[S])$.

  Proof: Let $H = H^+[E^+[S]]$. Any trip which leaves $\text{int } D^+[E^+[S]]$ crosses H. If $H = \emptyset$, then the proposition is proven. If not, then H must be non-compact. For, if H is compact, there are finitely many Alexandroff nbds. covering H, say $B_1, \dots, B_K$. There exists a pt. $p \in I^+[S] \smallsetminus D^+(E^+[S])$; suppose $p \in B_i$. Then there is a point $q_i \in I^+[S] \smallsetminus D^+(E^+[S])$

such that $q_{i_1} \ll p$ and $q_{i_1} \in B_{i_1} - B_i$ for

some i_1. The process continues by induction to yield

$p \gg q_{i_1} \gg q_{i_2} \gg \cdots \gg q_{i_n} \gg \cdots$. However, when the process has

been repeated K times two ⟨at least⟩ of the q's must lie in the

same B_{i_L}. By construction, the trip between these two must leave and

then reenter B_{i_L} contradicting the fact that $\overline{I^+[s]}$ is strongly

causal.

So H is non-compact. Let ξ be any nowhere vanishing timelike

vector field on M and construct the family of trajectories from

$E^+[s]$. If one of these trajectories leaves $\mathrm{int}\, D^+\!\left(E^+(s)\right)$,

it intersects H in one point. If they all leave, then a

homeomorphism is established between $E^+(s)$ and H, which is

impossible since $E^+(s)$ is compact, and H is not. So one

trajectory remains inside.

Theorem 5: The following are mutually inconsistent in any spacetime :

    (1) There are no closed trips

    (2) Every endless causal geodesic contains a pair of conjugate points

    (3) There is a future (past) - trapped set S in M.

Outline of proof: (see [5]). Under the assumption that all three

conditions are satisfied, (1) and (2) imply that M is strongly

causal. This implies the existence of a future endless geodesic

in $\mathrm{int}\, D^+\!\left(E^+(s)\right)$ say γ. Put $T = I^-(\gamma) \cap E^+(s)$. T is past-

trapped and so there exists α, a past endless causal geodesic in

$\mathrm{int}\, D^-\!\left(E^-(T)\right)$. Choose a sequence a_i receding into the past on α

without a limit point and a sequence c_i on γ going to the future.

The sets $J^-(c_i) \cap J^+(a_i)$ are compact, for all i, and so there

is a maximal geodesic μ_i in $\mathcal{C}(a_i, c_i)$ for each i. The

intersections with T ⟨compact⟩ have a limit point ξ and a limiting

causal direction. Construct the causal geodesic μ which has this direction at q. By (2) there is a pair of conjugate points u, v on μ. Assume $u < v$. Since conjugate points vary continuously with conditions defining the geodesic, there must be sequences $\{u_i\}$ and $\{v_i\}$, $u_i \rightarrow u$, $v_i \rightarrow v$, with v_i conjugate to u_i on μ_i. Since the $\{a_i\}$ and $\{c_i\}$ have no accumulation point, there must exist a K such that $a_K < u_K < v_K < c_K$. But this contradicts the maximality of the geodesic μ_K .

BIBLIOGRAPHY

1. Carter, B., Phys. Rev. <u>174</u>, 1559 (1968).

2. Geroch, R., Domain of dependence, J. Math. Phys. <u>11</u>, 437 (1970).

3. Hawking, S. W., The occurrence of singularities in cosmology,
       I Proc. Roy. Soc. A <u>294</u>, 511 (1966); II Proc. Roy. Soc. A
       <u>295</u>, 490 (1966); III Proc. Roy. Soc. A. <u>300</u>, 187 (1967).

4. Hawking, S. W., Singularities and the geometry of space-time (Adams Prize
       Essay, Cambridge U.), (1966).

5. Hawking, S. W., and Penrose, R. The singularities of gravitational collapse
       and cosmology, Proc. Roy. Soc A <u>314</u>, 529 (1970).

6. Hicks, N. J., <u>Notes</u> <u>on</u> <u>Differential</u> <u>Geometry</u>, Princeton: D. van Nostrand
       Inc. (1965).

7. Kronheimer, E. A., and Penrose, R., Proc. Camb. Phil. Soc. Loud. <u>63</u>,
       481 (1967).

8. Milnor, J., <u>Morse</u> <u>Theory</u>, Princeton: Princeton University Press (1968).

9. Penrose, R., An analysis of the structure of space-time
       (Adams Prize Essay, Cambridge U.), (1966).

10. Penrose, R., Article in <u>Battelle</u> <u>Rencontres,</u> <u>1967</u> <u>Lectures</u> <u>in</u>
       <u>Mathematics</u> <u>and</u> <u>Physics</u> (ed. DeWitt and Wheeler)
       New York: W. A. Benjamin, Inc.

11. Penrose, R., Techniques of Differential Topology in General Relativity,
       A.M.S. Colloquium Series (to be published).

A SIMPLE DERIVATION OF THE GENERAL REDSHIFT FORMULA

Dieter R. Brill

Yale University [*]
New Haven, Conn. 06520 [†]

In an arbitrary space-time, let two observers 1, 2 be located at events P_1, P_2 which can be connected by a null geodesic (see figure). Let U_1, U_2 be the 4-velocities of the observers, and let K be tangent to the null geodesic and covariant constant (i.e., K is transported parallel along the null geodesic). E. Schrödinger [1] has given a simple formula for the "frequency shift" between the two observers for any signal propagated along null geodesics:

$$r \equiv \frac{\nu_1}{\nu_2} = \frac{\langle U_1 , K \rangle}{\langle U_2 , K \rangle} . \tag{1}$$

This result clearly does not depend on the observers' being in the same coordinate patch, so that a derivation using global differential geometry is appropriate. The derivation below may be pedagogically useful, since it does not rely on integration or variational principles.

Let t_1 and t_2 be the proper times measured by the two observers. We assume that there are further null geodesics, crossing the worldline of the first observer at all t_1 in a finite interval of proper time surrounding P_1, which also reach the second observer -- i.e., that the first observer has a finite time in which to communicate with the second. These geodesics can be labelled either by the crossing time at 1 or at 2, and therefore define a function $t_2(t_1)$. The frequency

---

*Present address: Department of Physics and Astronomy, University of Maryland, College Park, Md. 20742.

†Supported in part by the National Science Foundation.

ratio r is simply the derivative of this function,

$$r = dt_2/dt_1 \ .$$
(2)

The null geodesics can always be reparametrized such that their affine parameter equals 0 on the worldline of the first observer, and it equals 1 on the worldline of the second observer. The points on the null geodesics can be labelled by the pair (t_1, s) and the derivatives

$$V = \partial/\partial t_1, \qquad K = \partial/\partial s$$
(3)

can be considered as tangent vectors in the space-time. We then have

$$V(s=0) = U_1 \ ;$$
(4)

however,

$$V(s=1) = (dt_2/dt_1) \ \partial/\partial t_2 = r U_2.$$
(5)

We now show that $\langle V, K \rangle$ is constant along the null geodesics, from which the result (1) follows immediately.

We need the condition for null geodesics,

$$\langle K, K \rangle = 0 \quad (6a) \ , \qquad D_K K = 0 \qquad (6b)$$

and the compatibility of the covariant differentiation D,

$$D_X Y - D_Y X = [X, Y] \qquad (7a) \ , \quad X \langle Y, Z \rangle = \langle D_X Y, Z \rangle + \langle Y, D_X Z \rangle \qquad (7b)$$

We apply (7b) to

$$\frac{\partial}{\partial s} \langle V, K \rangle = K \langle V, K \rangle = \langle D_K V, K \rangle$$

(where the second term vanishes in view of (6b)) and use $[V, K] = 0$ (due to Eq. (3)) and Eq. (7a) to rewrite this as

$$\frac{\partial}{\partial s} \langle V, K \rangle = \langle D_V K, K \rangle = \tfrac{1}{2} V \langle K, K \rangle = 0$$

47

in view of (6a). Thus $\langle V,K \rangle$ is indeed constant along null geodesics, and we have

$$\langle V,K \rangle_1 = \langle U_1,K \rangle = \langle V,K \rangle_2 = r \langle U_2,K \rangle , \tag{8}$$

which proves Eq. (1).

The evaluation of the right side of (1) is particularly simple if the observers' 4-velocities are parallel to a Killing or conformal Killing vector. For example, in a static space-time with metric

$$ds^2 = - e^{2\emptyset} dt^2 + d\sigma^2 ,$$

with $\emptyset$ and $d\sigma^2$ independent of t, $\partial/\partial t$ is a Killing vector, hence $\langle \partial/\partial t, K \rangle$ is constant along the null geodesics. The 4-velocity of static observers is the unit vector $U = e^{-\emptyset} \partial/\partial t$, hence

$$r = e^{\phi_2 - \phi_1} ,$$

the usual expression in terms of the difference of gravitational potential $\emptyset$.

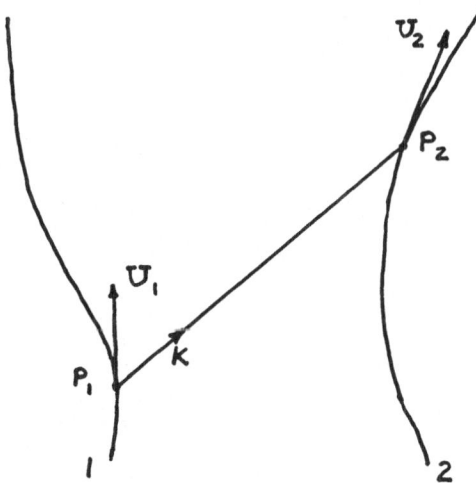

REFERENCE

1. E. Schrödinger, <u>Expanding Universes</u>, Cambridge University Press (1956), p. 49.

SOME REMARKS ON A RADIATING SOLUTION OF THE
EINSTEIN-MAXWELL EQUATIONS

M. Walker and W. Kinnersley*

Department of Physics, University of Texas at Austin

INTRODUCTION

Mathematically speaking, we might define a <u>spacetime</u> as a pair (M, g_{ab}), consisting of a 4-dimensional connected Hausdorff differentiable manifold M, together with a differentiable pseudo-Riemannian metric g_{ab} of signature $(+---)$ on M. [1] In General Relativity, one does not usually study <u>all</u> such objects but instead employs a criterion to select spacetimes of physical interest. In order to state one of Einstein's criteria, we require some notational conventions: ∇_a denotes the unique covariant derivative satisfying $\nabla_a g_{bc} = 0$. The Riemann tensor associated with ∇_a is defined by

$$2\nabla_{[a}\nabla_{b]} t_c = R_{abc}{}^d t_d$$

where t^a is an arbitrary vector field and square brackets denote antisymmetrization. From $R^a{}_{bcd}$ one obtains the Ricci tensor R_{ac} by contraction: $R_{ac} = R^b{}_{abc}$. The Riemann tensor may be invariantly split up into R_{ac}, $R = g^{ac}R_{ac}$, and a trace-free part C_{abcd} (called the Weyl tensor). Physically speaking, C_{abcd} describes the free gravitational field, while any matter present is described by R_{ab}. One case in which we are interested here is the absence of matter: $R_{ab} = 0$; in this case, $C_{abcd} = R_{abcd}$, and a solution (M, g_{ab}) to the equation $R_{ab} = 0$ is often called a <u>vacuum</u> solution. $R_{ab} = 0$ comprises Einstein's vacuum field equations.

---

*Present address: Aerospace Research Laboratories, Wright-Patterson Air Force Base, Ohio.

A solution to the (vacuum) Einstein-Maxwell equations is a little more complicated; it is a triple (M, g_{ab}, F_{ab}) where M and g_{ab} are as before, and F_{ab} is an antisymmetric tensor describing an electromagnetic field. F_{ab} is linked to g_{ab} via the electromagnetic stress-energy tensor

$$R_{ab} = F_{ac} F^c_{\ b} + \frac{1}{4} F_{ec} F^{ec} g_{ab},$$

and Maxwell's equations

$$\nabla_{[a} F_{bc]} = 0, \quad \nabla_a F^a_{\ b} = 0.$$

The last of these equations is interpreted as saying that there are no charges or currents present; F_{ab} is a free electromagnetic field. Although these equations appear complicated, in many cases when a solution to $R_{ab} = 0$ is known, an "electrified" version is quickly found.

In any case, there are many more solutions known than there are solutions which are understood, so we will utterly ignore the vast problem of how one goes about finding them. From now on when we speak of a "solution", we will mean a solution to the vacuum Einstein-Maxwell equations.

A solution is usually "given" by writing down the components, with respect to some coordinate system or field of basis vectors, of the metric g_{ab} and the electromagnetic field F_{ab}. One can then find the gravitational field C_{abcd} by differentiation.

Given a solution, one is faced with two related problems. Since g_{ab} is given only locally, a "largest" manifold M on which g_{ab} is defined should be found. (This will not be unique in general.) Secondly, one should find a physical interpretation of the given fields. We will carry out this program for a particularly interesting example. For more details, we refer the reader to our paper [2].

THE SOLUTION

The metric is given, in the above sense, by

$$ds^2 = (\tilde{x} + \tilde{y})^{-2}(F d\tilde{t}^2 - F^{-1} d\tilde{y}^2 - G^{-1} d\tilde{x}^2 - G d\tilde{z}^2) \tag{1}$$

where $G = G(\tilde{x}) = a + a_1\tilde{x} + a_2\tilde{x}^2 + a_3\tilde{x}^3 + a_4\tilde{x}^4$ and $F = F(\tilde{y}) = -G(-\tilde{y})$. The a_i are constants. We do not give the electromagnetic field, but merely remark that $F_{ab} = 0 \Longleftrightarrow a_4 = 0$. The case $a_4 = 0$ was found originally by T. Levi-Civita [3] fifty-two years ago, but its physical interpretation remained a mystery until one of us (W. K.) found the charged version recently [4].

The plan is to interpret physically the free parameters in Eq. (1) and then to investigate the manifold which admits (1) as a Lorentz metric. As a first step, note that not all of the a_i are significant since a transformation of the form

$$t = c_0\tilde{t}$$

$$y = Ac_0\tilde{y} + c_1$$

$$x = Ac_0\tilde{x} - c_1$$

$$z = c_0\tilde{z} \tag{2}$$

merely modifies the metric (1) by a constant conformal factor:

$$ds^2 = A^{-2}(x + y)^{-2}(F dt^2 - F^{-1} dy^2 - G^{-1} dx^2 - G dz^2) \tag{3}$$

while preserving its form. By a judicious choice of constants in Eq. (2), the five a_i may be reduced to three parameters. By demanding that Eq. (3) have certain limits as various of the parameters vanish, we chose to put G into the form

$$G = 1 - x^2 - 2mAx^3 - e^2A^2x^4$$

$$F = -G(-y), \tag{4}$$

the three parameters being m, e, and A. In order to assign physical meaning to these, we will show how the limiting procedure works, but first it is convenient to choose some new coordinates.

NEW COORDINATES

Our new coordinates are based on the fact that the Weyl tensor of the metric (1) is algebraically special. The algebraic classification scheme referred to may be briefly described as follows. The Weyl tensor determines four null directions at each point, called <u>principal null directions</u>. If these are distinct at generic points, then C_{abcd} is said to be algebraically <u>general</u>. If, on the other hand, certain of the principal null directions are degenerate, then C_{abcd} is said to be <u>special</u>. In the present case, the principal null directions coincide in pairs: C_{abcd} is said to be of type $\{2,2\}$. Moreover, the null congruences associated with these null directions are geodesic, non-shearing, hyper-surface orthogonal and diverging: we denote the affinely parametrized null tangent vector to one such congruence by ℓ^a, and to the other n^a, where $\ell^a n_a = 1$. ℓ^a and n^a are called (repeated) <u>principal null vectors</u>.

The above conditions on ℓ^a and n^a imply that there exist scalars u and v on the spacetime such that

$$\ell_a = \nabla_a u,$$

and

$$n_a = \nabla_a v.$$

Let r be an affine parameter along one of the congruences. Then we may choose as coordinates either $\{u, r, x, z\}$ or $\{v, r, x, z\}$.

In terms of the coordinates $\{t, y, x, z\}$, u, v, and r are given by

$$Au = t + \int^y F^{-1} dy,$$

$$Av = t - \int^y F^{-1} dy,$$

$$Ar = (x + y)^{-1}.$$

Note that the null hypersurfaces $N^+(u)$ defined by $u = $ const., and $N^-(v)$ defined by $v = $ const. are intrinsically singled out by the geometry of our spacetime.

What is more, each of $N^+(u)$ and $N^-(v)$ admits a natural foliation into spacelike 2-surfaces $\Sigma^+(u, r)$ given by u = const., r = const., and $\Sigma^-(v, r)$ given by v = const., r = const., respectively.

In terms of the "null coordinates" defined above, the metric (3) becomes

$$ds^2 = \begin{cases} Hdu^2 + 2dudr + 2Ar^2dudx - r^2(G^{-1}dx^2 + Gdz^2), \\ Hdv^2 - 2dvdr - 2Ar^2dvdx - r^2(G^{-1}dx^2 + Gdz^2), \end{cases} \quad (5)$$

where

$$H = -A^2r^2G + ArG' + (1 + 6mAx + 6e^2A^2x^2)$$
$$- 2(m + 2e^2Ax)r^{-1} + e^2r^{-2}.$$

INTERPRETATION

We will return to the scalar H later on, but first note that when $A = 0$, Eq. (5) reduces to the well-known Reissner-Nordstrøm solution for the field exterior to a spherically symmetric charged body of mass m and charge e, expressed in retarded (u) or advanced (v) null coordinates. Hence, we are led to a tentative identification of the parameters m and e, although as we shall see later, they are not the true mass and charge when $A \neq 0$.

When $m = e = 0$, the Weyl tensor vanishes and the spacetime is consequently flat. With $G = 1 - x^2$, $x = \cos \theta$ and $z = \phi$, the first of Eqs. (5) becomes

$$ds^2 = (1 - 2Ar \cos \theta - A^2r^2 \sin^2\theta)du^2 + 2dudr$$
$$- 2Ar^2 \sin \theta \, dud\theta - r^2(d\theta^2 + \sin^2\theta \, d\phi^2). \quad (6)$$

It turns out that the coordinates $\{u, r, \theta, \phi\}$ are adapted to a timelike hyperbola W in Minkowski space, or world line of constant acceleration A. This may be seen by transforming to Cartesian coordinates $\{\bar{t}, \bar{x}, \bar{y}, \bar{z}\}$ in Minkowski space as follows:

$$\bar{t} = (A^{-1} - r \cos \theta)\sinh Au + r \cosh Au,$$
$$\bar{z} = (A^{-1} - r \cos \theta)\cosh Au + r \sinh Au,$$

$$\bar{x} = r \sin \theta \cos \phi,$$

$$\bar{y} = r \sin \theta \sin \phi.$$

The locus $r = 0$ is the uniformly accelerating world line W:

$$\left. \begin{array}{l} \bar{t} = A^{-1} \sinh Au \\ \bar{z} = A^{-1} \cosh Au \\ \bar{x} = \bar{y} = 0 . \end{array} \right\} \quad \bar{z}^2 - \bar{t}^2 = A^{-2}.$$

Eliminating θ and ϕ from the above transformation, we find

$$(\bar{t} - A^{-1} \sinh Au)^2 - (\bar{z} - A^{-1} \cosh Au)^2 - \bar{x}^2 - \bar{y}^2 = 0,$$

which tells us that the $N^+(u)$ are retarded null cones with their vertices on the world line. Similarly, the $N^-(v)$ are advanced null cones based on the world line. The 2-surfaces $\Sigma^+(u, r)$ are spherical sections of the null cones $N^+(u)$, or wavefronts for the fields emanating from $r = 0$. θ and ϕ are spherical coordinates on these wavefronts.

Notice also that the world line $r = 0$ divides the $(\bar{t}, \bar{z})$ plane into two regions corresponding to $\theta = 0$ and $\phi = \pi$. $\theta = 0$ we call the North Polar region and $\theta = \pi$ the South Polar region. This distinction, although artificial in flat space, becomes essential in curved space. The situation is illustrated in Fig. 1. Observe that, with the sole exception of the ray along the South Pole, all of the null geodesic generators [5] of $N^+(u)$ eventually cross the null hyperplane $\mathcal{H}_0$. A final observation on the flat space situation is that there are actually two hyperbolae present here, the second one being obtained by reflecting the $(\bar{t}, \bar{z})$ plane about $\bar{z} = 0$. The coordinates $\{u, r, \theta, \phi\}$ actually only cover half of the space (and hence only one world line), the other half requiring a second coordinate patch. In the curved-space case, even more coordinate patches are required to cover the whole spacetime.

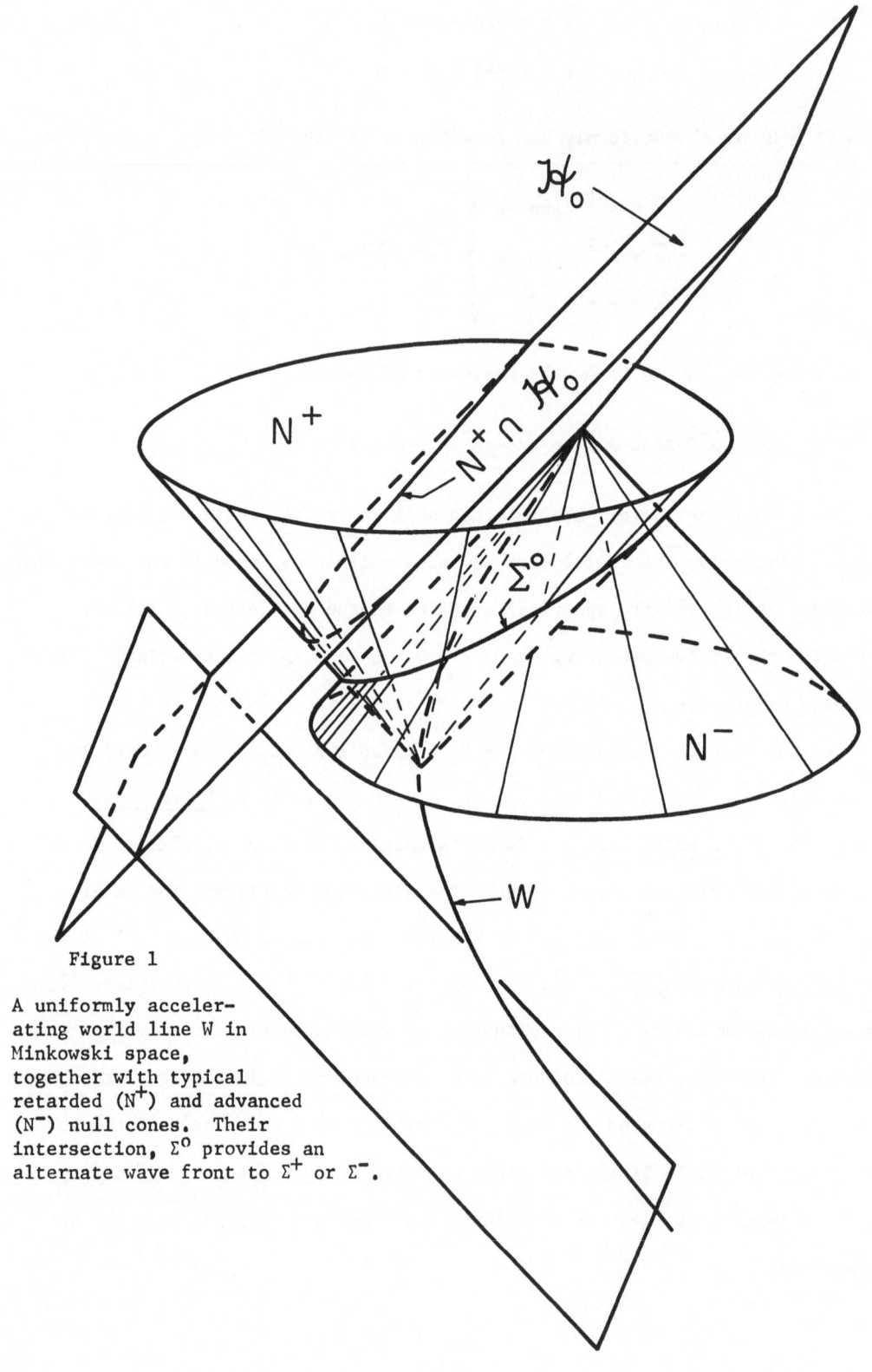

Figure 1

A uniformly acceler-
ating world line W in
Minkowski space,
together with typical
retarded (N^+) and advanced
(N^-) null cones. Their
intersection, Σ^0 provides an
alternate wave front to Σ^+ or Σ^-.

SYMMETRIES

Although there are many Killing vectors (generators of isometries) in Minkowski space, the form (6) of the flat space metric focuses attention on only two. These are $\xi^a = \delta^a{}_0 (\partial/\partial u)$, which is a boost or uniform acceleration transformation in the $(\bar{t}, \bar{z})$ plane, and $\zeta^a = \delta^a{}_3 (\partial/\partial \phi)$, a rotation in the $(\bar{x}, \bar{y})$ plane. The important point for our purposes is that the function H occurring in Eq. (5) is the norm of ξ^a:

$$\xi^a \xi_a = H$$
$$= 1 - 2Ar \cos\theta - A^2 r^2 \sin^2\theta$$
$$= A^2(\bar{z}^2 - \bar{t}^2), \qquad \text{(in flat space)}$$

and is consequently an invariant, conveying information about the geometry of spacetime. We have plotted H as a function of r and θ in Fig. 2. The null hyperplane $\mathcal{H}_0$ of Fig. 1 appears as $H = 0$ in Fig. 2, and the earlier remark concerning the South Polar ray is made clear: H increases linearly along this ray but passes through zero along all others.

CURVED SPACE

We now pass to a study of the curved space case which will parallel the above discussion. Let us study the null hypersurface $N^+(u)$ by investigating the 2-surfaces $\Sigma^+(u, r)$ which slice it up. The metric (5) restricted to Σ^+ is

$$d\sigma^2 = -r^2(G^{-1}dx^2 + Gdz^2). \tag{7}$$

Recall that $G = 1 - x^2 - 2mAx^3 - e^2 A^2 x^4$. A first step is to decide on the ranges of the coordinates x and z. The range of x is determined by requiring that $G \geq 0$ (so that the signature of the metric is $(+ - - -)$) and that Σ^+ be compact (it was a sphere in flat space). A glance at G reveals that there exist $x_2 < 0 < x_1$ such that $G(x) > 0$ for $x_2 < x < x_1$ and $G(x_1) = G(x_2) = 0$. (In flat space, $x_2 = -x_1 = -1$.)

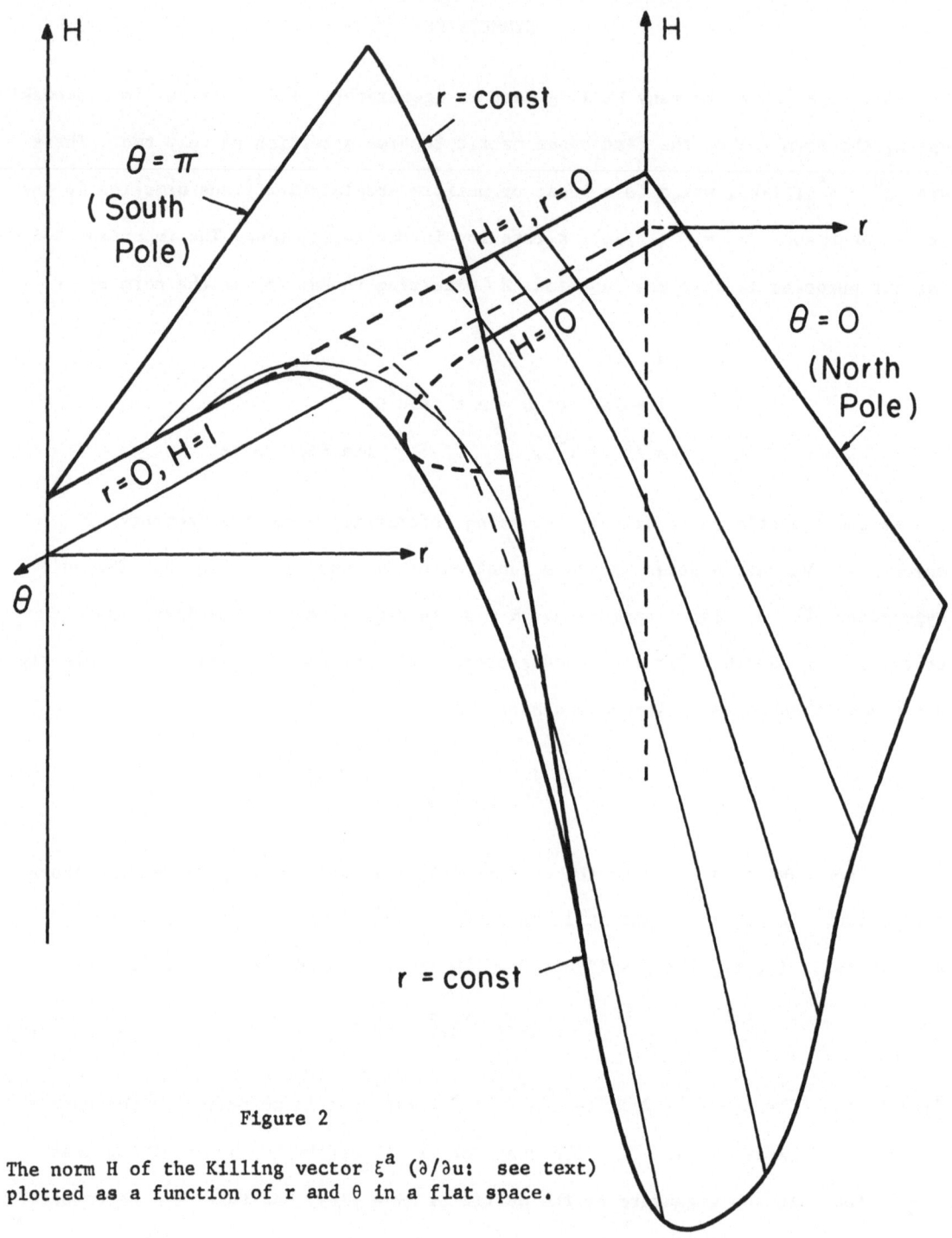

Figure 2

The norm H of the Killing vector ξ^a ($\partial/\partial u$: see text) plotted as a function of r and θ in a flat space.

THE NODE

In order to proceed further, let us embed Σ^+ as a surface of revolution in Euclidean space. To do this, define

$$\theta = \int_{x_2}^{x_1} G^{-1/2} dx,$$

$$\phi = \kappa z \quad \text{with} \quad \kappa = -\frac{1}{2} \frac{dG}{dx}\bigg|_{x_1},$$

$$\rho(\theta) = \kappa^{-1}\left[G(x(\theta))\right]^{1/2}.$$

Then Eq. (7) becomes

$$d\sigma^2 = -r^2(d\theta^2 + \rho^2(\theta)d\phi^2),$$

where $0 \leq \theta \leq \theta_2 = \int_{x_2}^{x_1} G^{-1/2} dx$, $0 \leq \phi \leq 2\pi$. We have chosen κ so that Σ^+ is smooth at the North Pole $\theta = 0$, as is shown in Fig. 3. We do this, however, at the expense of the South Pole $\theta = \theta_2$. We may define a measure $\varepsilon \equiv \varepsilon(\theta_2) = -\frac{d\rho}{d\theta}\bigg|_{\theta_2}$ of the "conicality" of Σ^+ at the South Pole: Σ^+ is smooth there only if $\varepsilon = 1$. If $\varepsilon \neq 1$, then Σ^+ has a nodal singularity at the South Pole similar to the vertex of a cone. Fig. 4 shows ε as a function of eA and mA, and we see that ε is unity only when $mA = 0$ or when $eA = mA > 12^{-1/2}$.

Hence, except in these two cases, N^+ inherits a nodal line along the South Pole. The situation is more serious than this, however; the spacetime itself has a nodal singularity in the form of a timelike 2-surface extending from the world line out to infinity in the South Polar direction. Excepting the case $eA = mA > 12^{-1/2}$, we may even assign a physical meaning to the node. We have interpreted our solution as representing a uniformly accelerating charged mass. Now when $mA \neq 0$, we expect from classical mechanics that a force of magnitude mA acting in the South Polar direction is required to produce the acceleration. We suggest that the nodal singularity is a manifestation of this force. The fact that the node disappears when $m = 0$ supports this conclusion, but we are as yet unable to account for the disappearance of the node when $mA = eA > 12^{-1/2}$.

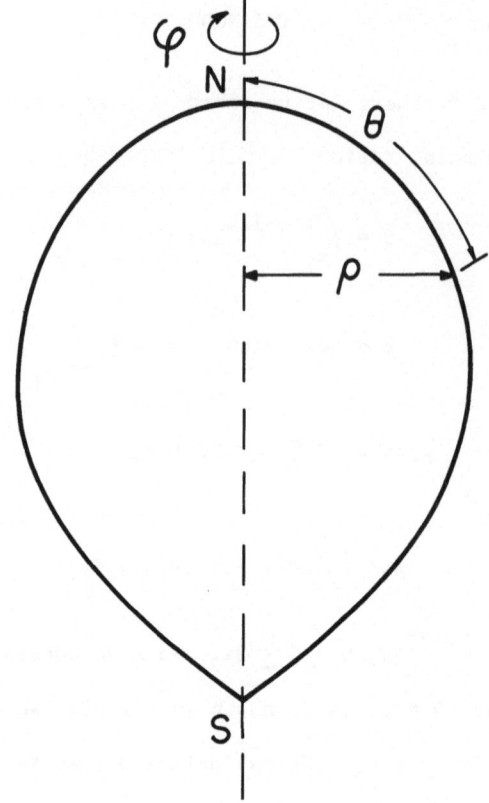

Figure 3

The embedding diagram for the spacelike 2-surface Σ^+. A typical "teardrop" is shown.

59

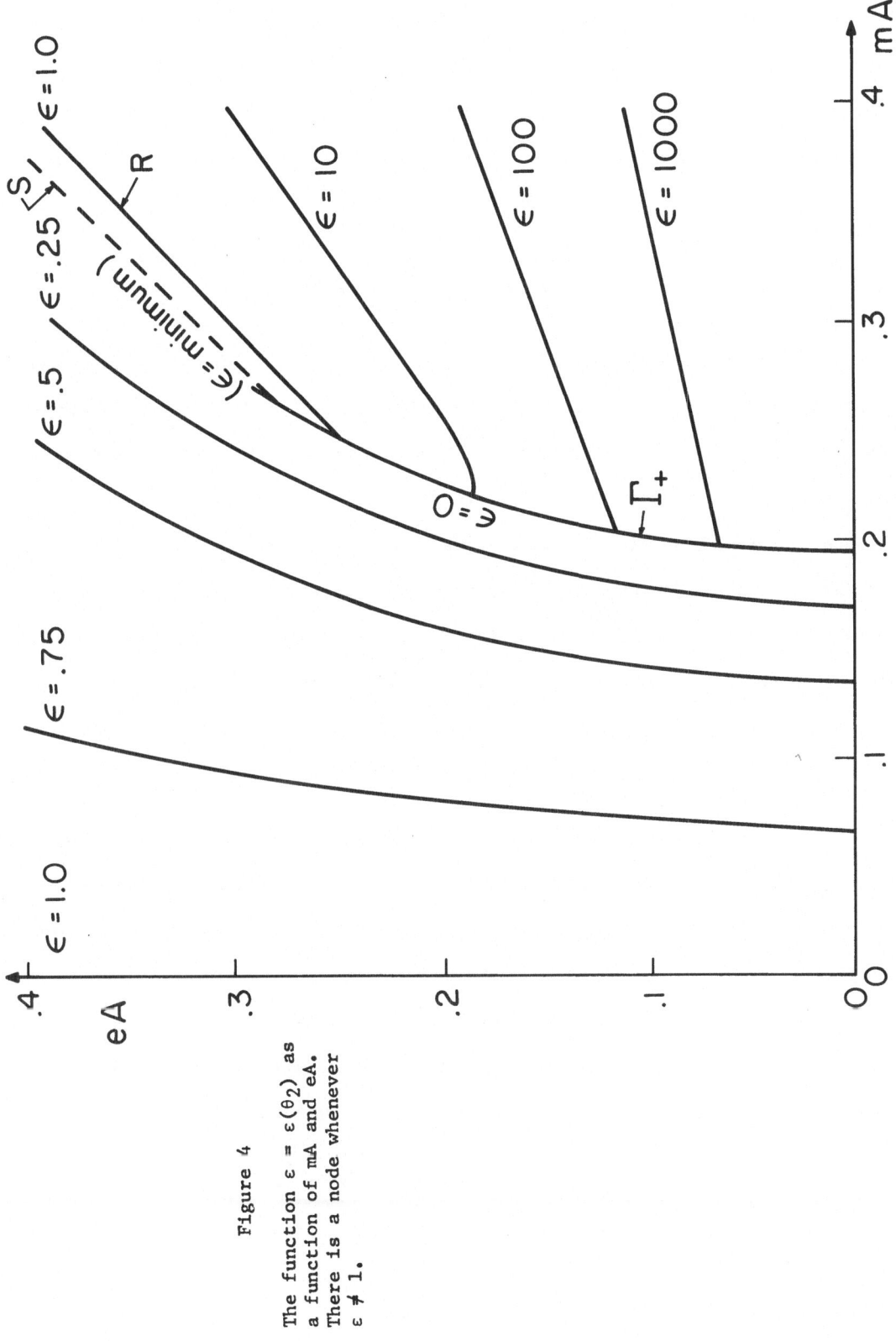

Figure 4

The function ε = ε(θ₂) as a function of mA and eA. There is a node whenever ε ≠ 1.

We make two final remarks about the node. The first is that the curvature is apparently well behaved there. The second is that since our spacetime has <u>no</u> unique tangent space at a nodal point it is <u>not</u> a differentiable manifold in the strict sense. Instead we have a differentiable manifold with singularities of the type discussed by R. F. Reynolds at this conference.

CLASSIFICATION OF SOLUTIONS

As the parameters m, A, and e change, the surfaces $\Sigma^+(u, r)$ also change in appearance. In fact, we can classify the solutions in our 3-parameter family by the shape of Σ^+. This is shown in Fig. 5. The idea is that the number and type of real roots of the equation $G = 0$ govern the shape of Σ^+. Above the line Γ_-, the surface Σ^+ resembles Fig. 3, which we have called a teardrop. Between Γ_- and Γ_+, there are actually two allowed ranges for x in which $G \geq 0$. On Γ_+, Σ^+ has an infinite tail, while to the right of Γ_+, Σ^+ recontracts bringing a second lobe with it to resemble a dumbbell. On R, the dumbbell has no node.

NULL GEODESICS

We come now to the global properties of our spacetime: what does the manifold M look like? We remark first of all that none of the coordinate systems exhibited so far covers all of M; that is to say, there exist geodesics which run off any of our patches within a finite affine distance from any point inside the patch. This is easiest to see from the null geodesics since these may be displayed explicitly. Let $k^a = (du/d\lambda, dr/d\lambda, dx/d\lambda, dz/d\lambda)$ be the tangent vector to an affinely parametrized null geodesic: $k^c \nabla_c k^a = 0$ and $k^c k_c = 0$. We find that

$$k^a = (H^{-1}[E + R], -R - AP, r^{-2}P, r^{-2}G^{-1}J_z),$$

where

$$R = \pm(E^2 - r^{-2}HJ^2)^{1/2}, \tag{8}$$

and

61

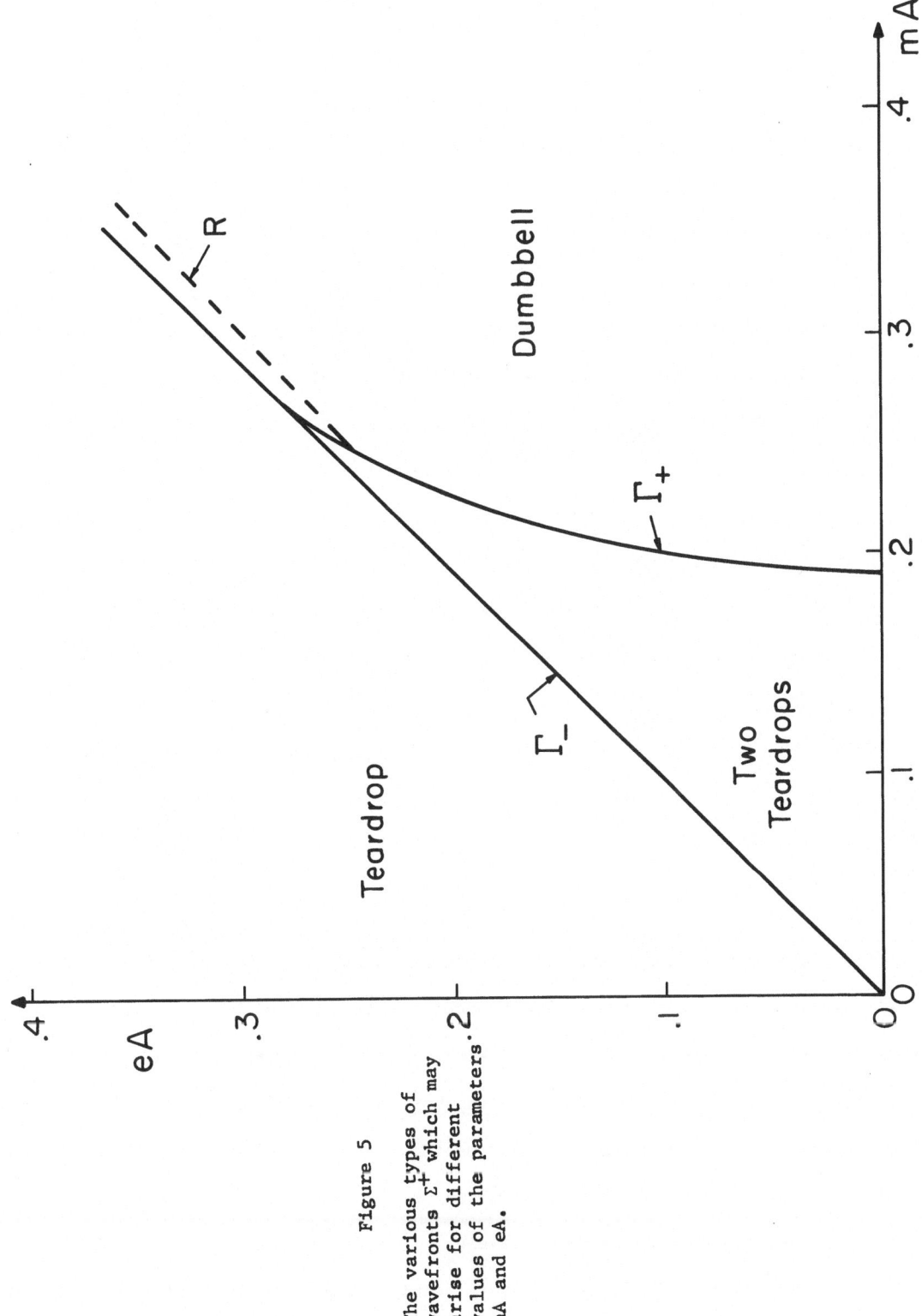

Figure 5

The various types of
wavefronts Σ^+ which may
arise for different
values of the parameters
mA and eA.

$$P = \pm(GJ^2 - J_z^2)^{1/2}.$$

Here, E, J, and J_z are constants of the motion on any one geodesic.

From Eq. (8), we see that a null geodesic can leave the $\{u, r, x, z\}$ patch in one of three ways:

(a) One of the components of k^a may become infinite. This cannot happen to the z-component since P is real and $G \rightarrow 0$ requires $J_z = 0$. The u-component becomes infinite when $H = 0$, and in this case the geodesic passes to another patch. We will return to this point in a moment.

(b) The coordinate r may go to zero within a finite affine distance. The reality of P and R in this case requires $J = J_z = 0$. Such a geodesic is incomplete, but since it encounters infinite curvature at $r = 0$, the geodesic is inextendible.

(c) If $J_z = 0$, a geodesic may cross the node. The patchwork required here is just the usual spherical kind and the node presents no difficulty in this connection.

Consequently, the only null geodesics which are incomplete and extendible are those along which H may become zero. Before proceeding to extend these geodesics, let us discuss the hypersurfaces $H = 0$.

HORIZONS

H is the norm of a hypersurface-orthogonal Killing vector ξ^a which does not itself vanish when H does; ξ^a merely becomes null. Hence, $H = 0$ defines null hypersurfaces generated by the Killing vector ξ^a, which is null there. These stationary hypersurfaces are called Killing horizons [6] and in the present case are of two types. To understand them, it is useful to look at Figs. 6 and 7 where H is plotted against r and x for the cases $e = 0$ and $emA \neq 0$ with the function G having three and four zeros respectively.

Fig. 6 represents H for a uniformly accelerating Schwarzschild solution, and as we might expect, there are two horizons. The outer one $\mathcal{H}_o$ (cf. Fig. 2) is the analogue of the null hyperplane occurring in flat space in connection with uniformly

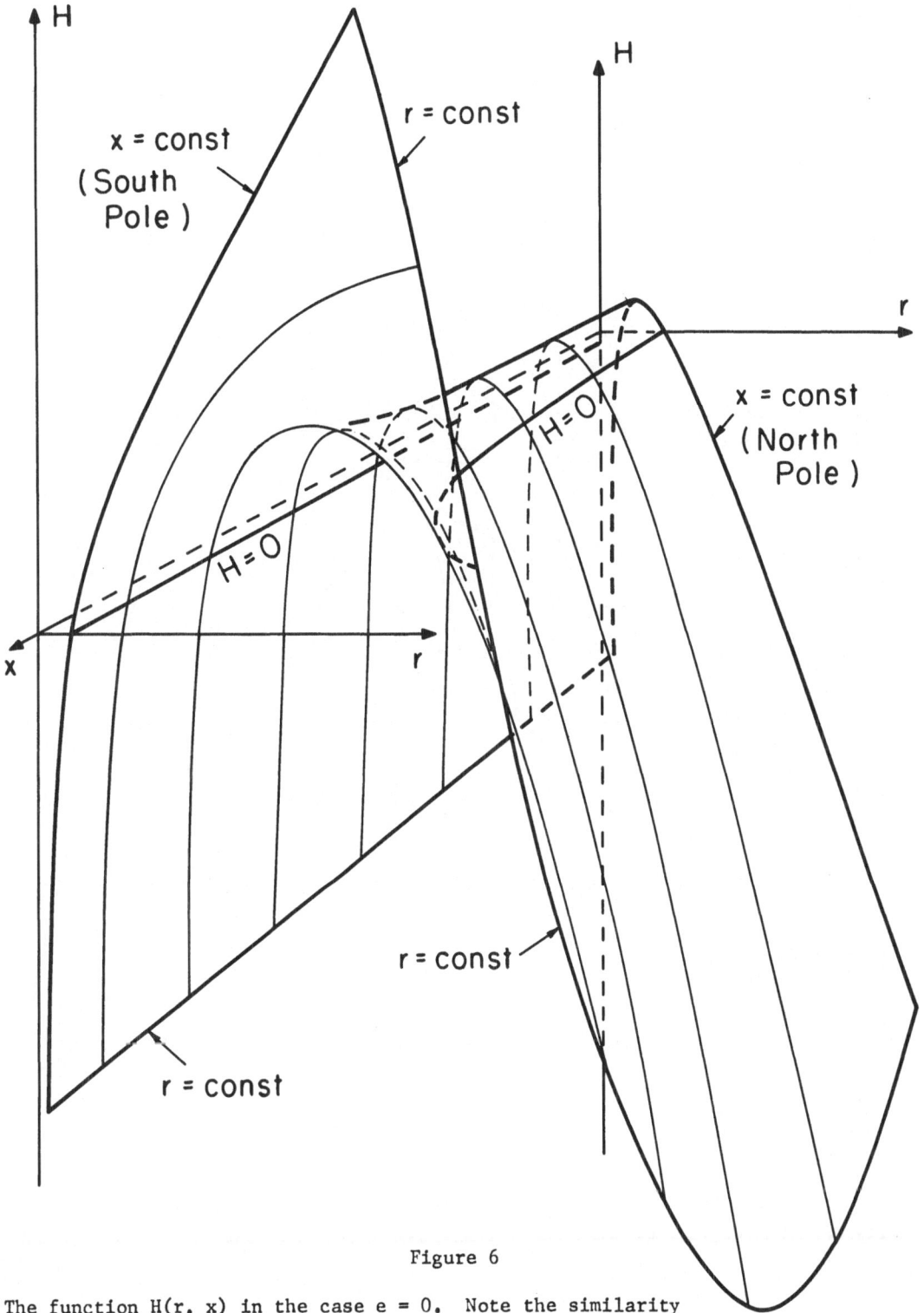

Figure 6

The function H(r, x) in the case e = 0. Note the similarity
with Fig. 2, as well as the appearance of an inner event horizon.

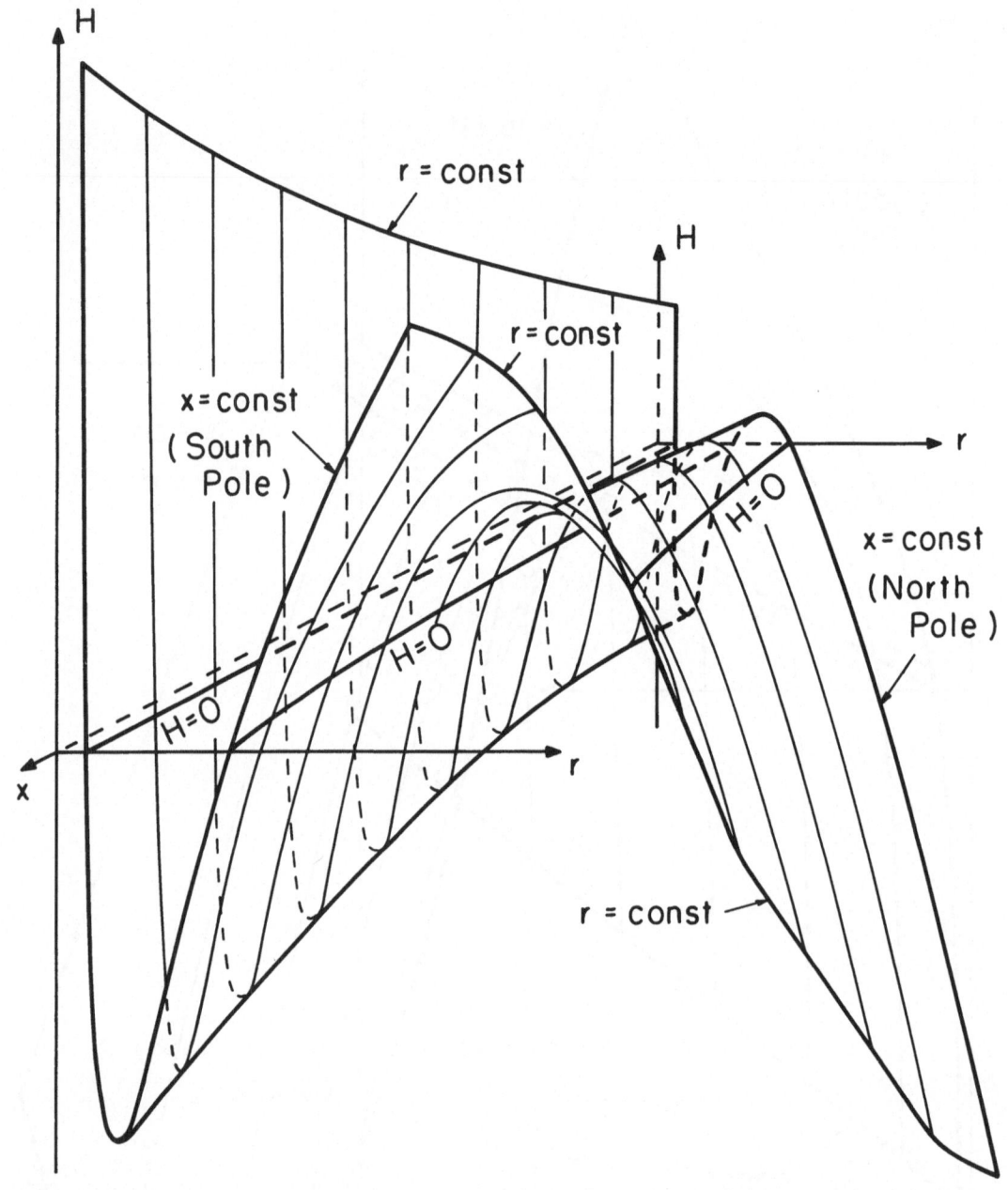

Figure 7

The function H(r, x) for the charged case when G has four zeros. Again note the
similarity with Fig. 2 for large r, and the presence this time of two inner event
horizons.

accelerated motion. The inner horizon $\mathcal{H}_1$ surrounds the world line r = 0 and is analogous to the _event_ _horizon_ of the Schwarzschild solution: it represents the boundary of the exterior region from which observers traveling along timelike curves may escape to infinity. Once an observer crosses $\mathcal{H}_1$, he is inevitably annihilated by the infinite tidal forces at r = 0.

Fig. 7 represents H for a uniformly accelerating Reissner-Nordstrøm solution (with e < m). Here there are two inner horizons, an anti-event as well as an event horizon, together with the outer horizon due to the accelerated motion. (Again, compare Fig. 2.) This situation is shown schematically in Fig. 8.

EXTENSIONS

Having decided that H = 0 defines respectable null hypersurfaces in our spacetime, we return to the question of extending the null geodesics across them. For convenience, consider those geodesics with $J = J_z = 0$. This implies that x and z are constant along the geodesic, and that we may restrict our attention to the timelike 2-surfaces x = const., z = const. These surfaces T(x, z) are actually invariantly defined, since they contain the repeated principal null vectors ℓ^a and n^a. We call T(x, z) the _principal_ _timelike_ _2-surfaces_. It is particularly convenient that we are only required to study 2-dimensional timelike surfaces since there is a well-known prescription for both carrying out the extension and giving a schematic diagram of the result [6].

Before giving the extensions, however, let us ask what the T(x, z) surfaces are in flat space. From the transformation equations given earlier, we find that T(x, z) is given by

$$\bar{t}^2 - \bar{z}^2 - (\bar{x} \sec \theta + A^{-1} \cot \theta)^2 = -A^{-2} \csc^2 \theta. \tag{9}$$

Equation (9) is the equation of a ruled elliptical hyperboloid which contains the world lines r = 0 and is actually generated by the principal null congruences. In Fig. 9(a), we show one such surface for $\theta = \pi/2, \phi = \pi/2$. In this case, T is actually a hyperboloid of revolution. In Fig. 9(b), we show the $\bar{t} = 0$ sections of

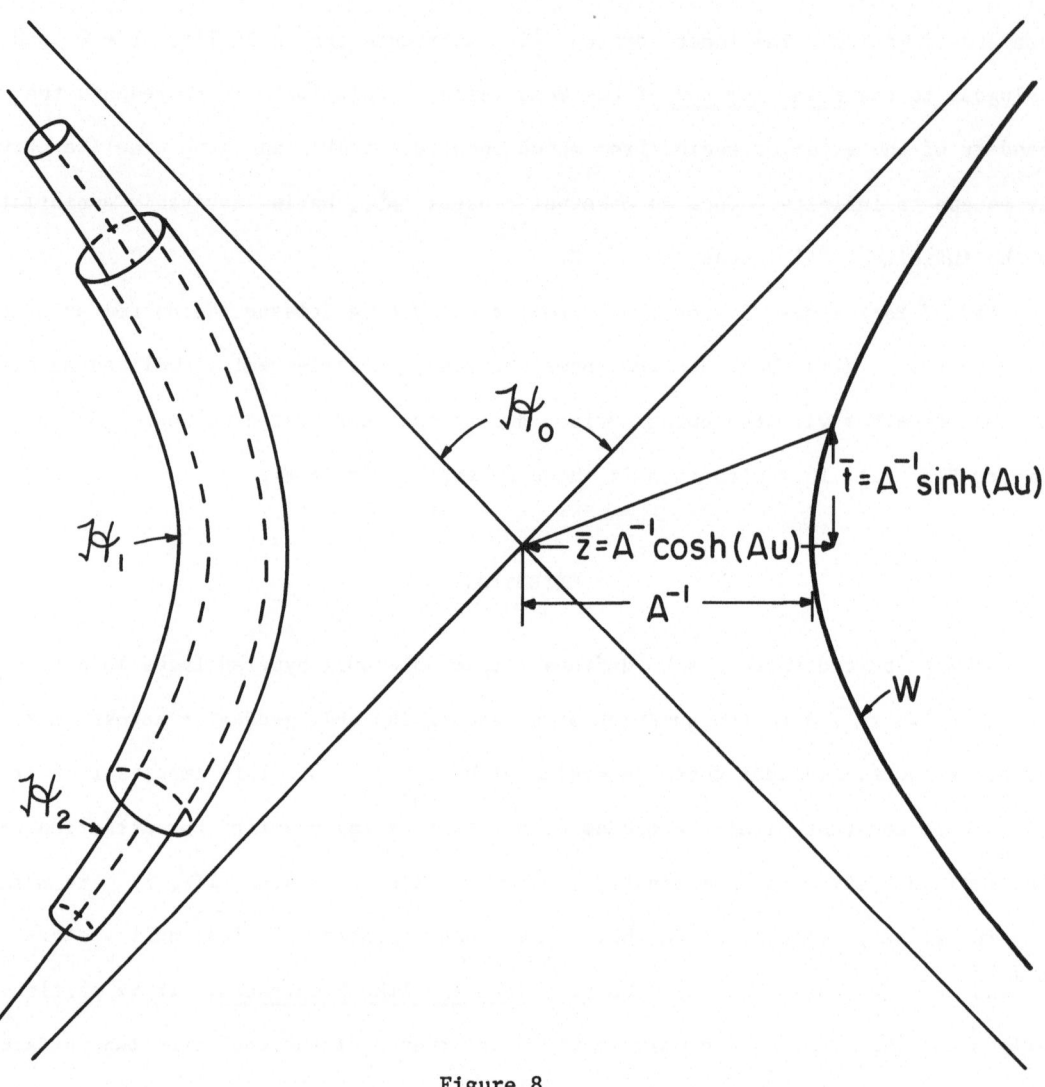

Figure 8

A schematic representation of the situation of Fig. 7. The world line W in flat space is also shown, together with its parametrization in terms of u.

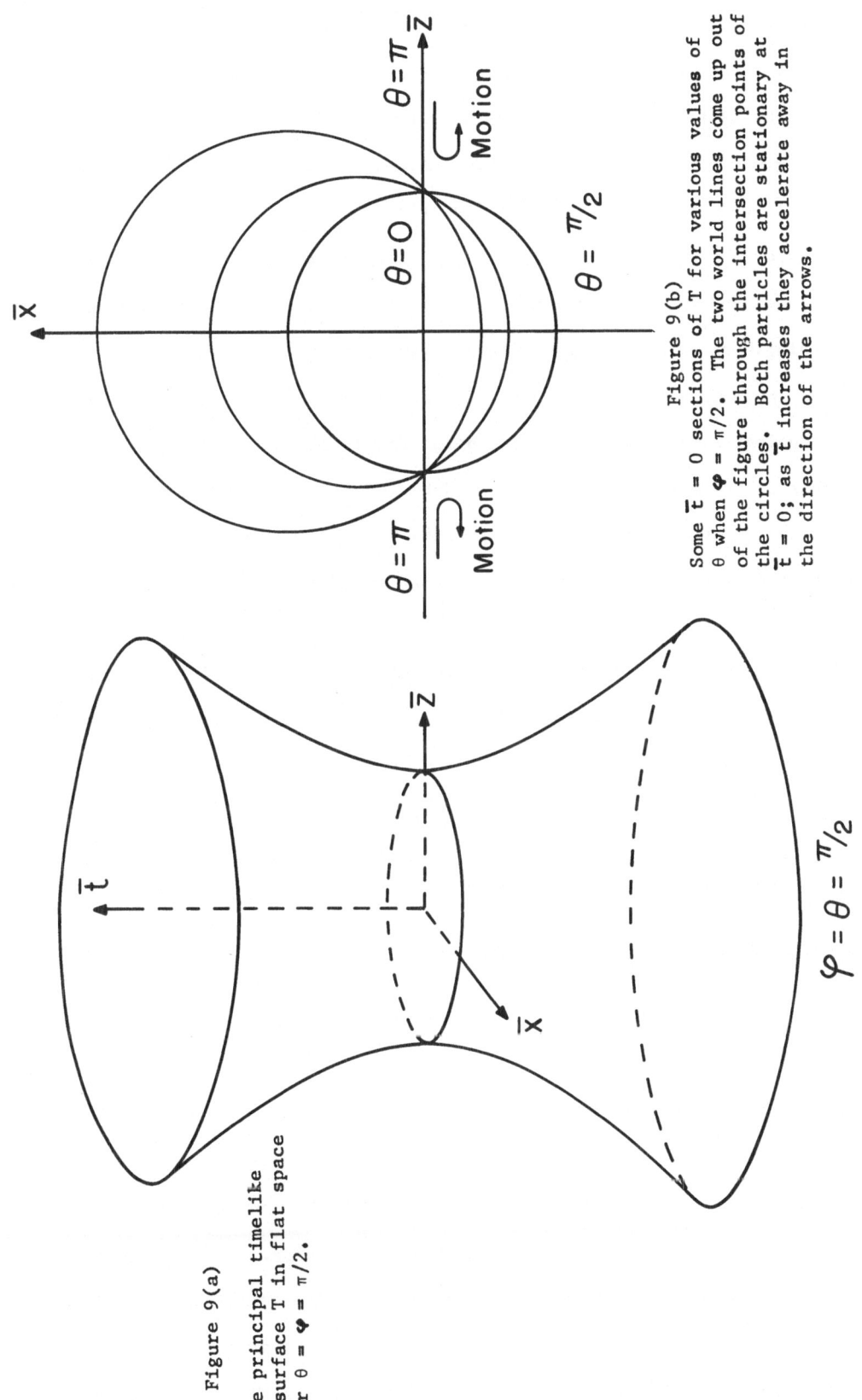

Figure 9(a)

The principal timelike
2-surface T in flat space
for $\theta = \varphi = \pi/2$.

$\varphi = \theta = \pi/2$

Figure 9(b)

Some $\bar{t} = 0$ sections of T for various values of
θ when $\varphi = \pi/2$. The two world lines come up out
of the figure through the intersection points of
the circles. Both particles are stationary at
$\bar{t} = 0$; as $\bar{t}$ increases they accelerate away in
the direction of the arrows.

T for various values of θ when φ = π/2. T degenerates into the North Polar plane between the two world lines when θ = 0 and to two disjoint South Polar planes extending from the world lines to infinity when θ = π.

Making use of the block diagram technique [7], the T(x, z) surfaces may be extended, on noting that the spacetime metric restricted to T is $Hdu^2 + 2dudr$. The horizons H = 0 appear as lines at $45°$; conformal infinity is represented by double lines, and the singularity at r = 0 by a jagged line. Each block represents a single coordinate patch. Examples in which G has two, three, and four zeros are shown in Figs. 10, 11, and 12, respectively. In Fig. 13, we show Fig. 11 with a topological identification performed which shows explicitly the non-uniqueness of the extensions obtained. More accurately, the extension of T should show two copies of each figure, glued together along their singularities to bring out the fact that, in flat space, T is a hyperboloid.

The solutions discussed here have <u>trapped surfaces.</u> In fact, the surfaces $\Sigma^+(u, r)$ are trapped inside the inner horizon $\mathscr{R}_1$ when e = 0, and between the two inner horizons when emA ≠ 0 and G has four zeros: all of the null geodesics meeting Σ^+ orthogonally in these regions converge into the future.

Just as the Schwarzschild solution has null geodesics which spiral around just outside the horizon, at r = 3m, so our solution also has spiraling null geodesics. They occur for example at r = 3m when x = 0.

ASYMPTOTIC BEHAVIOR

We would like to turn now to a study of the asymptotic properties of the C-metric. We would like to clarify in what sense the solution may be regarded as asymptotically flat, and to what extent it fulfills the long-range ambition of exhibiting an exact radiation field from a bounded source. We will continue to conduct parallel investigations on two levels: the linearized theory and the full theory. The reason for this approach is that results in simple closed form are possible only in the linearized solution. However it is then not difficult to show

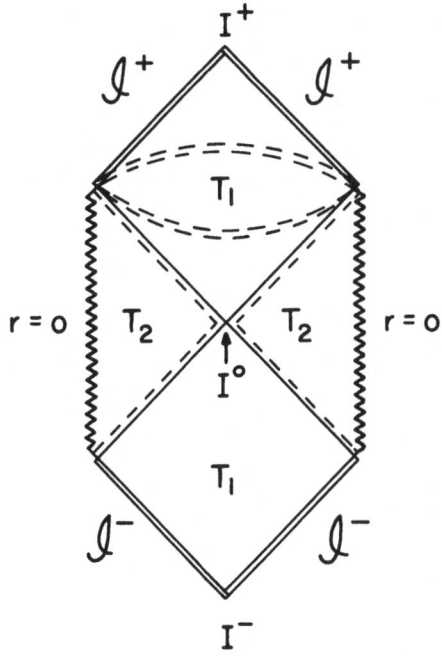

Figure 10

Block diagram for the "Teardrop" region of Fig. 5.

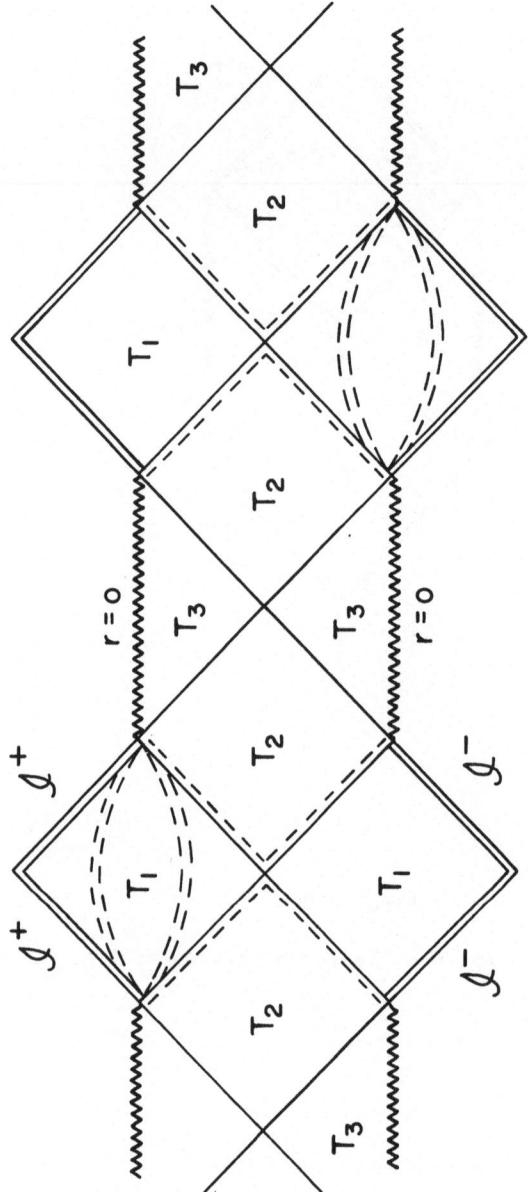

Figure 11

Block diagram for the "Two Teardrops" region of Fig. 5 when G has three zeros.

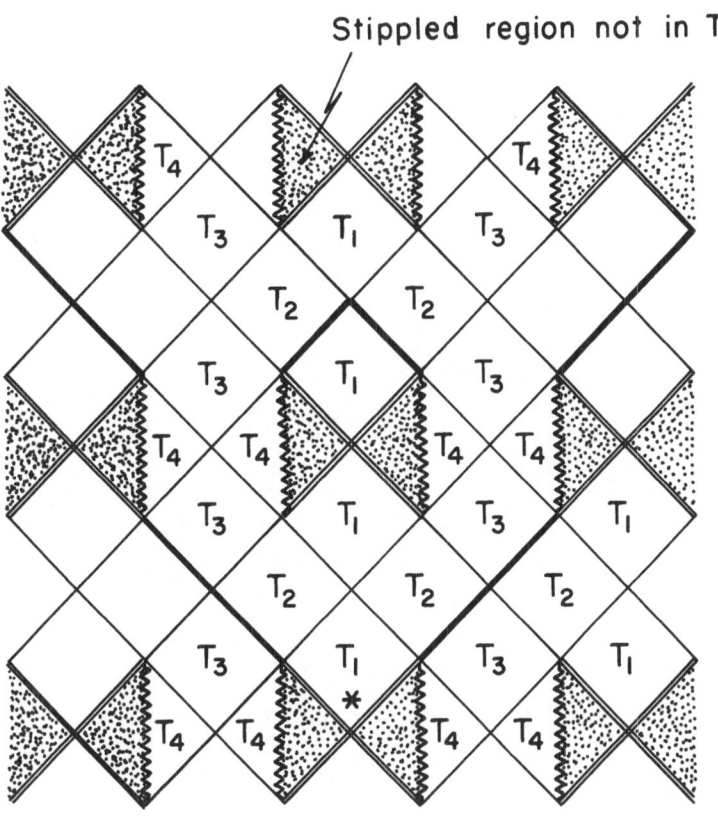

Figure 12

Block diagram for the "Two Teardrops" region of Fig. 5 when G has four zeros.

72

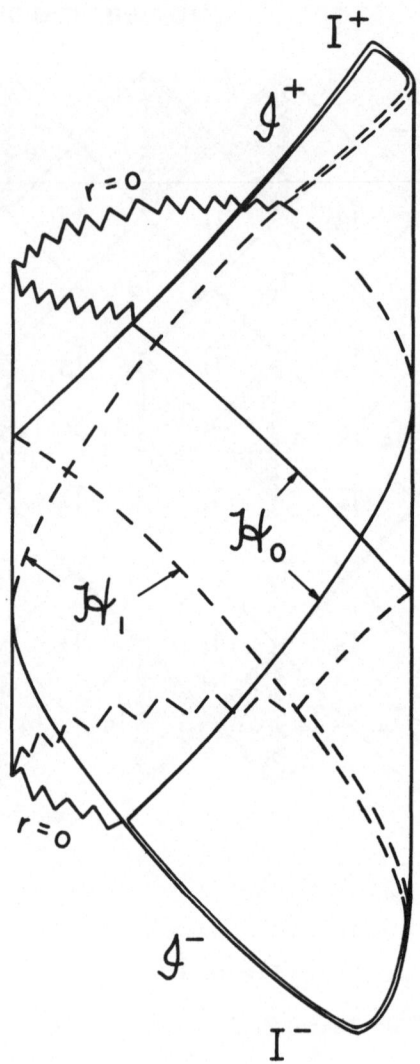

Figure 13

Fig. 11 "wrapped up."

that most of the features are qualitatively unchanged when we consider the full
theory.

In order to completely describe the curvature we introduce a complex spacelike
unit vector m^a orthogonal to the principal null directions. Thus ℓ^a, n^a, m^a form
a complex orthonormal tetrad basis. The projected components of the Weyl and
Maxwell tensors are defined as follows;

$$\Psi_0 = -C_{abcd} \, \ell^a m^b \ell^c m^d \, ,$$

$$\Psi_1 = -C_{abcd} \, \ell^a m^b \ell^c n^d \, ,$$

$$\Psi_2 = -C_{abcd} \, \ell^a m^b \bar{m}^c n^d \, ,$$

$$\Psi_3 = -C_{abcd} \, \ell^a n^b \bar{m}^c n^d \, ,$$

$$\Psi_4 = -C_{abcd} \, \bar{m}^a n^b \bar{m}^c n^d \; ; \tag{10}$$

$$\phi_0 = F_{ab} \ell^a m^b \, ,$$

$$\phi_1 = \tfrac{1}{2} F_{ab} (\ell^a n^b + \bar{m}^a m^b) \, ,$$

$$\phi_2 = F_{ab} \bar{m}^a n^b , \tag{11}$$

these being the only ones which are linearly independent. At each point therefore,
a general gravitational field in a general basis is described by five complex
numbers and an electromagnetic field by three. However due to the fact that the
C-metric is a type $\{2, 2\}$ solution, and since we have chosen a basis which is
aligned with the principal null direction, at present the only components which are
in fact nonzero are Ψ_2 and ϕ_1. For their values we obtain:

$$\Psi_2 = - (m + 2e^2 Ax) r^{-3} + e^2 r^{-4} ,$$

$$\phi_1 = \tfrac{1}{2} \sqrt{2} \, er^{-2} . \tag{12}$$

These quantities provide measures of the gravitational and electromagnetic field
strengths at each point, and they are effectively invariants. As expected they

both diverge at the worldline of the particle, r = 0, and they both tend to zero as r tends to infinity. At least in that sense the field is asymptotically flat. The next question to be settled is what this flat region r = ∞ looks like; that is, whether it is truly infinite in extent and whether or not it completely surrounds the "source" at r = 0. In the linearized theory the answers are clearly affirmative since in that case r measures affine distance away from the worldline W. In the full theory we have pointed out that for almost all choices of the parameters m, e, and A the surfaces Σ^+ where r = const. are still closed (topologically spherical) and that they lie on the outgoing null cones N^+ in such a way that as r increases, each Σ^+ completely encloses all previous ones.

NEWMAN-UNTI COORDINATES

The most direct way to convince ourselves that the spacetime is asymptotically flat in the usual sense would be to find a coordinate system in which the metric becomes Minkowskian in the limit of large r. This is essentially what we will do, except that instead of asymptotically Minkowskian coordinates we will seek ones which approach null retarded spherical coordinates, since they will be better suited to the description of outgoing radiation. The coordinate conditions to be imposed are ones suggested by Newman and Unti [8]:

$$
g^{ab} = \begin{Bmatrix}
0 & 1 & 0 & 0 \\
1 & 1 + O(\bar{r}^{-1}) & O(\bar{r}^{-3}) & 0 \\
0 & O(\bar{r}^{-3}) & \bar{r}^{-2} + O(\bar{r}^{-3}) & 0 \\
0 & 0 & 0 & \bar{r}^{-2} \csc^2 \theta + O(\bar{r}^{-3})
\end{Bmatrix} \quad (13)
$$

where the new coordinates are listed in the order u, r, θ, φ. Since Newman-Unti coordinates cannot usually be obtained in closed form, the metric is to be calculated as the first few terms in an asymptotic expansion in powers of $\bar{r}^{-1}$. Such an expansion will only be valid outside some radius of convergence $\bar{r} = \bar{r}_0(\theta, \phi)$. It is a temptation to suppose that if Newman-Unti coordinates exist then the space is asymptotically flat, but this is not the case. Counterexamples are provided by NUT space [9] and, unfortunately, the C-metric. What goes wrong in both of these instances is that the terms in the metric of order $\bar{r}^{-3}$ have poles in the angular variables. Thus in certain directions $\bar{r}_0$ becomes infinite and the asymptotic expansion fails to converge in any neighborhood of infinity. Even when such a situation arises one may still be able to talk about the existence of radiation, provided the Newman-Unti apparatus is used with care.

We seek a coordinate transformation of the form

$$\bar{u} = U + O(r^{-1}) \, ,$$

$$\bar{r} = Rr + O(1) \, ,$$

$$\bar{x} \equiv \cos\bar{\theta} = X + O(r^{-1}) \, ,$$

$$\bar{\phi} = \phi \, , \tag{14}$$

where U, R, X are the functions of u and x to be determined. When these forms along with the C-metric of Eq. (5) are substituted into the transformation equation,

$$g^{\bar{a}\bar{b}} = \frac{\partial x^{\bar{a}}}{\partial x^a} \frac{\partial x^{\bar{b}}}{\partial x^b} g^{ab} \, , \tag{15}$$

and the Newman-Unti conditions are imposed, we are left with a set of differential equations obeyed by U, R, and X:

$$U_u + AGU_x = R^{-1} \, ,$$

$$R_u + AGR_x = \tfrac{1}{2}AG_x R \, ,$$

$$X_u + AGX_x = 0 \, ,$$

$$R^2(1 - X^2) = \kappa^{-2} G \, ,$$

$$R^{-2}(1 - X^2) = GX_x^2 \, . \tag{16}$$

The subscripts here are used to denote partial derivatives. These equations have the general solution

$$U = \kappa A^{-1} \operatorname{sech}\chi \int_0^\chi G^{-3/2} + \alpha(\chi) \; ,$$

$$R = \kappa^{-1} G^{\frac{1}{2}}\cosh\chi \; ,$$

$$X = -\tanh\chi \; , \tag{17}$$

where

$$\chi = \kappa A_u - \kappa \int_0^\chi G^{-1} + C \; ; \tag{18}$$

C is an arbitrary constant, and $\alpha(\chi)$ an arbitrary function.

This freedom, or ambiguity, encountered in defining Newman-Unti coordinates is the well-known Bondi-Metzner-Sachs group [10]. The constant C represents a Lorentz boost along the axis of symmetry and we arbitrarily choose it to be zero. The function $\alpha(\chi)$ is the supertranslational freedom. By proper choice of α we may set U = 0 on any initial null hypersurface we desire. A desirable approach is to make U = 0 coincide with the principal null cone u = 0. This constraint can be used in principle to solve explicitly for α, although the resulting expression would be exceedingly complex.

If the above coordinate transformation were to be continued out to the next power of r^{-1}, we could obtain the r^{-3} terms in the metric which carry information about the radiation. Proceeding in this way eventually the news function, the Bondi mass and all the multipole moments would appear directly in the Newman-Unti metric tensor and need merely be read from there. A better way of obtaining these quantities, however, is by means of the Newman-Unti tetrad. We must have a new set of basis vectors tied to $\bar{u}$, $\bar{r}$, in the same ways that ℓ^a, n^a are tied to u, r. Namely, we will have

$$\bar{\ell}^a = \frac{\partial x^a}{\partial \bar{r}} \; ,$$

$$\bar{n}^a = \frac{\partial x^a}{\partial \bar{u}} - \tfrac{1}{2}\frac{\partial x^a}{\partial \bar{r}} + 0(r^{-1}). \tag{19}$$

The Lorentz transformation required to rotate from ℓ, n, m to $\bar{\ell}$, $\bar{n}$, $\bar{m}$ will induce transformations of the Ψ_A and ϕ_A according to the $(2,0)$ and $(1,0)$ representations of the Lorentz group. Hence in the new basis all of their components will be nonzero.

LINEARIZED RESULTS

At this point it is helpful to turn to the linearized theory where the situation can be more easily pictured. In the Minkowski diagram of Fig. 14 we have again shown the uniformly accelerating worldline W which is the locus $r = 0$, along with the null cones u = const. based upon it. Our Newman-Unti coordinates in flat space reduce to null spherical coordinates. Their origin $\bar{r} = 0$ is a straight worldline W' tangent to W. We have also shown some typical null cones $\bar{u}$ = const. Note that u = 0 and $\bar{u}$ = 0 are the same cone. At any point in the spacetime the principal null vectors ℓ and n will be proportional to the two unique null vectors connecting that point to the worldline W (one advanced and one retarded). The Newman-Unti basis vectors at the same point are proportional to the null connection vectors to W'. Without belaboring any further calculational details we now exhibit the complete linearized gravitational field in the new basis and new coordinates:

$$\Psi_0 = -\tfrac{3}{4} m A^2 \bar{u}^4 R^{-5} \sin^2 \bar{\theta} \, ,$$

$$\Psi_1 = \tfrac{3}{2} \sqrt{2} \, m A \bar{u}^2 R^{-5} \sin \theta \{ \bar{r} + \tfrac{1}{2} A \bar{u} \bar{v} \cos \bar{\theta} \} \, ,$$

$$\Psi_2 = -m R^{-5} \{ \bar{r}^2 + A \bar{u} \bar{v} \bar{r} \cos \theta + \tfrac{1}{4} A^2 \bar{u}^2 \bar{v}^2 P_2(\cos \bar{\theta}) \} \, ,$$

$$\Psi_3 = -\tfrac{3}{4} \sqrt{2} \, m A \bar{v}^2 R^{-5} \sin \bar{\theta} \, \{ \bar{r} + \tfrac{1}{2} A \bar{u} \bar{v} \cos \bar{\theta} \} \, ,$$

$$\Psi_4 = -\tfrac{3}{16} m A^2 \bar{v}^4 R^{-5} \sin^2 \bar{\theta} \, , \tag{20}$$

where

$$R = (\bar{r}^2 + A \bar{u} \bar{v} \bar{r} \cos \bar{\theta} + \tfrac{1}{4} A^2 \bar{u}^2 \bar{v}^2)^{\frac{1}{2}} \, ,$$

$$\bar{v} = \bar{u} + 2\bar{r} \, . \tag{21}$$

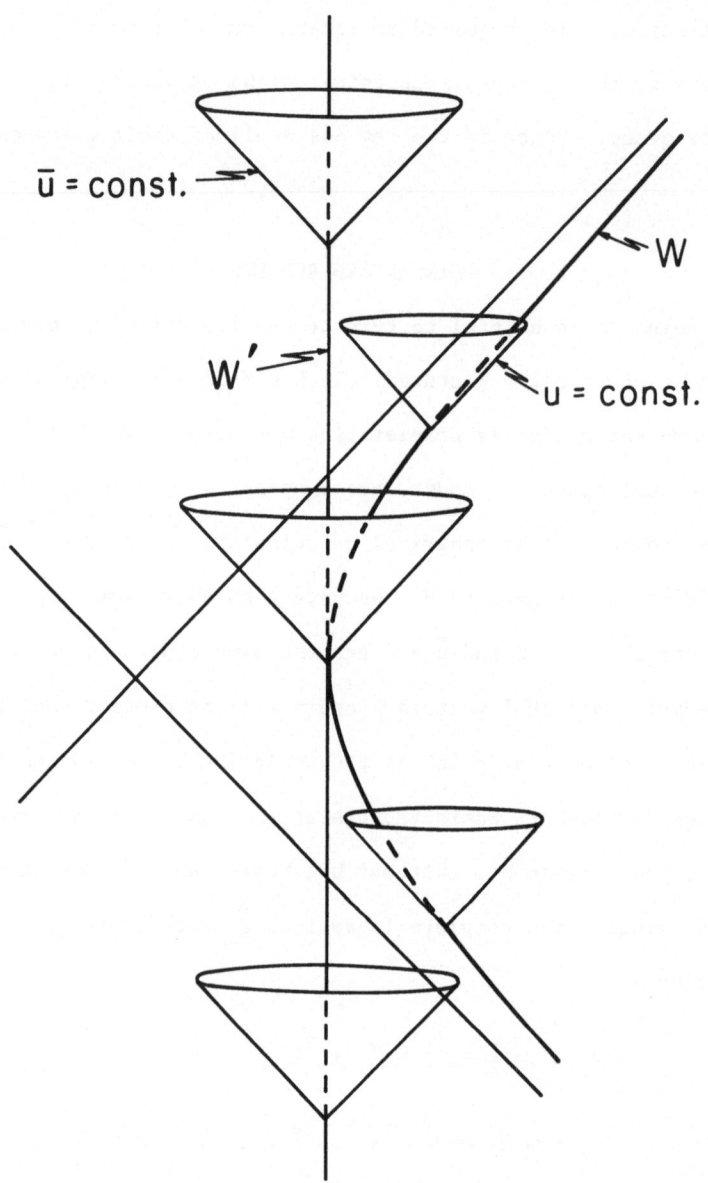

Figure 14

Newman-Unti coordinate system used for the linearized field.

If we hold $\bar{u}$ = const. and let $\bar{r} \to \infty$ we see that Ψ_4 is proportional to r^{-1}. This is sufficient to confirm the presence of outgoing radiation. The method of analyzing a linearized field by means of a multipole expansion has been given by Janis and Newman [11]. To find their multipole moments we must expand Ψ_0 in a series involving $P_\ell^2 (\cos\bar\theta)$. The generating function for the P_ℓ^2's is

$$\frac{3 c^2 (1 - x^2)}{(1 - 2cx + c^2)^{5/2}} = \sum c^\ell P_\ell^2 (x) \qquad (22)$$

which converges for $|c| < 1$. Applying this to the expression for Ψ_0 with

$$c = -\tfrac{1}{2} A^{-1} \bar{u} \bar{v} \bar{r}^{-1} , \qquad (23)$$

we obtain

$$\Psi_0 = \sum -(-2)^{-\ell} m A^\ell \bar{u}^{\ell+2} (\bar{u} + 2\bar{r})^{\ell-2} \bar{r}^{-\ell-3} P_\ell^2 (\cos\bar\theta) . \qquad (24)$$

The Janis-Newman moment q_ℓ is defined as the coefficient of $\bar{r}^{-\ell-3} P_\ell^2$ in Ψ_0. Hence

$$q_\ell = (-2)^{-\ell} m A^\ell \bar{u}^{2\ell} \quad \text{for} \quad |\bar{u}| < A^{-1} . \qquad (25)$$

For large $\bar{u}$ we may again use the generating function but this time with

$$c = -2A \bar{u}^{-1} \bar{v}^{-1} \bar{r} . \qquad (26)$$

We get

$$\Psi_0 = \sum (-2)^{\ell+1} m A^{-\ell-1} \bar{u}^{-\ell+1} (\bar{u} + 2\bar{r})^{-\ell-3} \bar{r}^{\ell-2} P_\ell^2 (\cos\bar\theta),$$

$$q_\ell = \frac{(-2)^{-\ell} (2\ell)!}{(\ell-2)! (\ell+2)!} m A^{-\ell-1} \bar{u}^{-1} \quad \text{for} \quad |\bar{u}| > A^{-1} . \qquad (27)$$

Although Ψ_0 was a continuous function, all the multipole moments are discontinuous at $\bar{u} = A^{-1}$. Fig. 15 shows why this might have been expected. Here we have a Penrose compactification [12] of the spacetime, so it is as if we are on the outside looking in. The asymptotic region appears as an outer surface consisting of

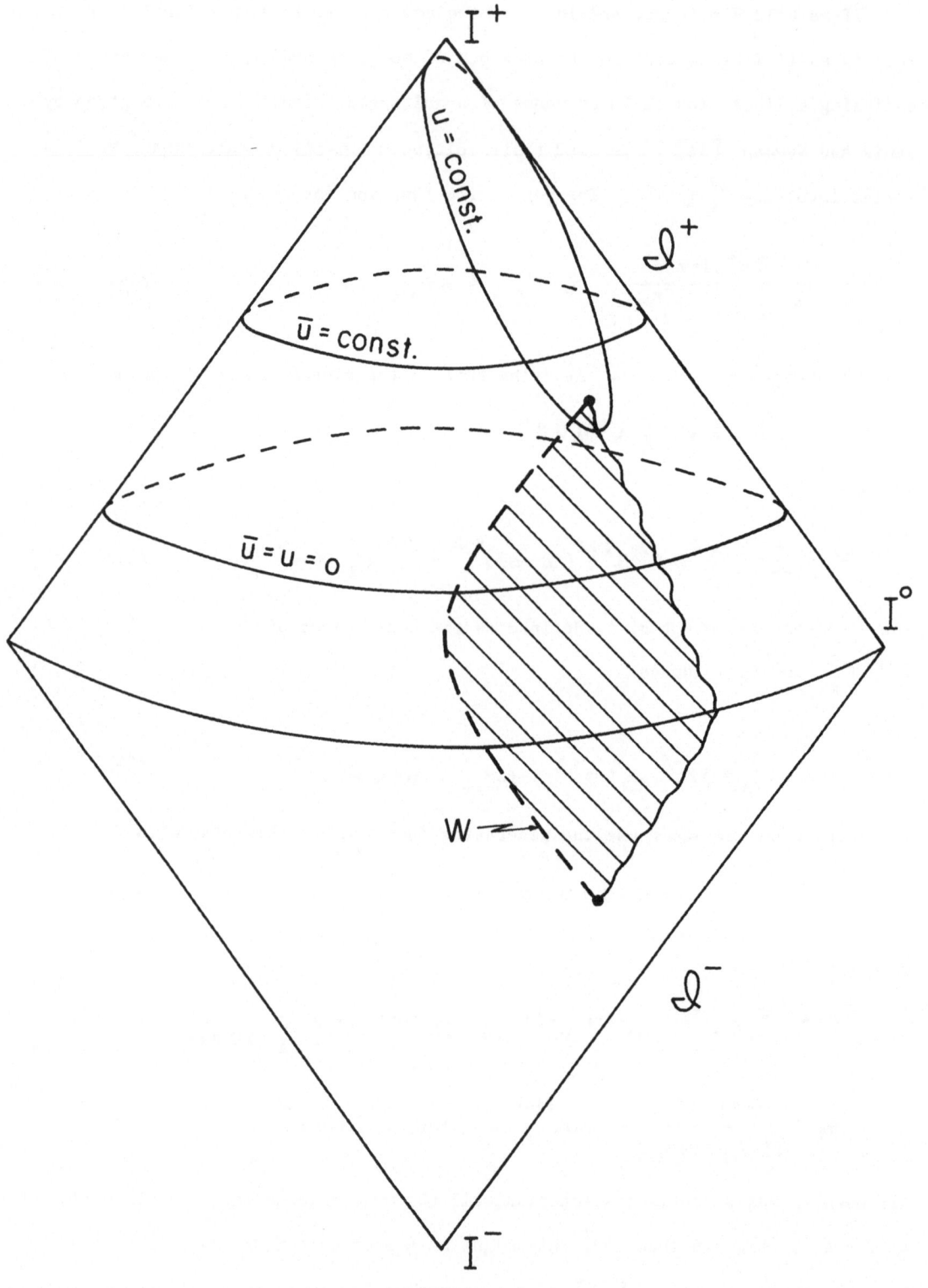

Figure 15

The compactified C-metric.

two null cones $\mathcal{J}^{\pm}$. The worldline W pierces $\mathcal{J}$ in two "bullet holes" or isolated singular points as shown. The nodal two-surface has been cross-hatched to show it also extends to meet infinity. The two families of outgoing null cones intersect $\mathcal{J}^+$ in conic sections, the difference between them being that all of the principal null cones u = const. loop about the bullet hole, whereas the u = const. sections are parallel ones. For $|u| < A^{-1}$ they enclose the bullet hole while for $|u| > A^{-1}$ they do not. In the first region the field will include a contribution from the particle itself but in the second the moments are describing only a pure radiation field. Thus the linearized mass q_0 is equal to m in region 1 and zero in region 2. Another quantity of particular interest is the Newman-Penrose constant [13], defined as the coefficient of $r^{-6}P_2^2$ in Ψ_0. From our expansion we see it is indeed time independent in both regions. However it, too, undergoes a jump discontinuity at $u = A^{-1}$.

EXACT RESULTS

We would like to do as complete an analysis for the exact solution, but this is not feasible. We have limited ourselves to calculating the monopole moment and the time derivative of the dipole moment. Together these transform under the Bondi-Metzner-Sachs group as the time and space components of a 4-vector called the Bondi momentum vector. Its definition is

$$P_0 = \lim_{\bar{r} \to \infty} \frac{1}{4\pi} \int \bar{r}^3 \, \Psi_2 \, d^2\bar{\Omega} \, ,$$

$$P_z = \lim_{\bar{r} \to \infty} \frac{3}{4\pi} \int \bar{r}^3 \, \Psi_2 \, P_1(\cos\bar{\theta}) \, d^2\bar{\Omega} \, , \qquad (28)$$

where the limits are taken holding u constant. The true electric charge is obtained as

$$E = \lim_{r \to \infty} \frac{1}{4\pi} \int \bar{r}^2 \varphi_1 \, d^2\bar{\Omega} \, . \qquad (29)$$

We further restrict ourselves to evaluating these quantities only on the principal null cone $\bar{u} = 0$, which is a shear-free surface. (On a null hypersurface with shear there would be an extra term in P_0 and P_z.) Even so we must resort to numerical calculation. Our results are displayed in Figs 16 and 17. We have plotted the invariants E and $M^2 = P_0^2 - P_z^2$ as functions of the two dimensionless parameters mA and eA. Near the origin where the linearized theory is valid, m and e agree well with the true mass and charge. However this is clearly not the case elsewhere, and in fact we see that $\vec{P}$ becomes null and even spacelike in certain regions.

For later times we expect $\vec{P}$ and E to show discontinuities at $\bar{u} = A^{-1}$ as in the linear case. In addition $\vec{P}$ will undergo a continuous secular decrease due to the mass loss from escaping radiation.

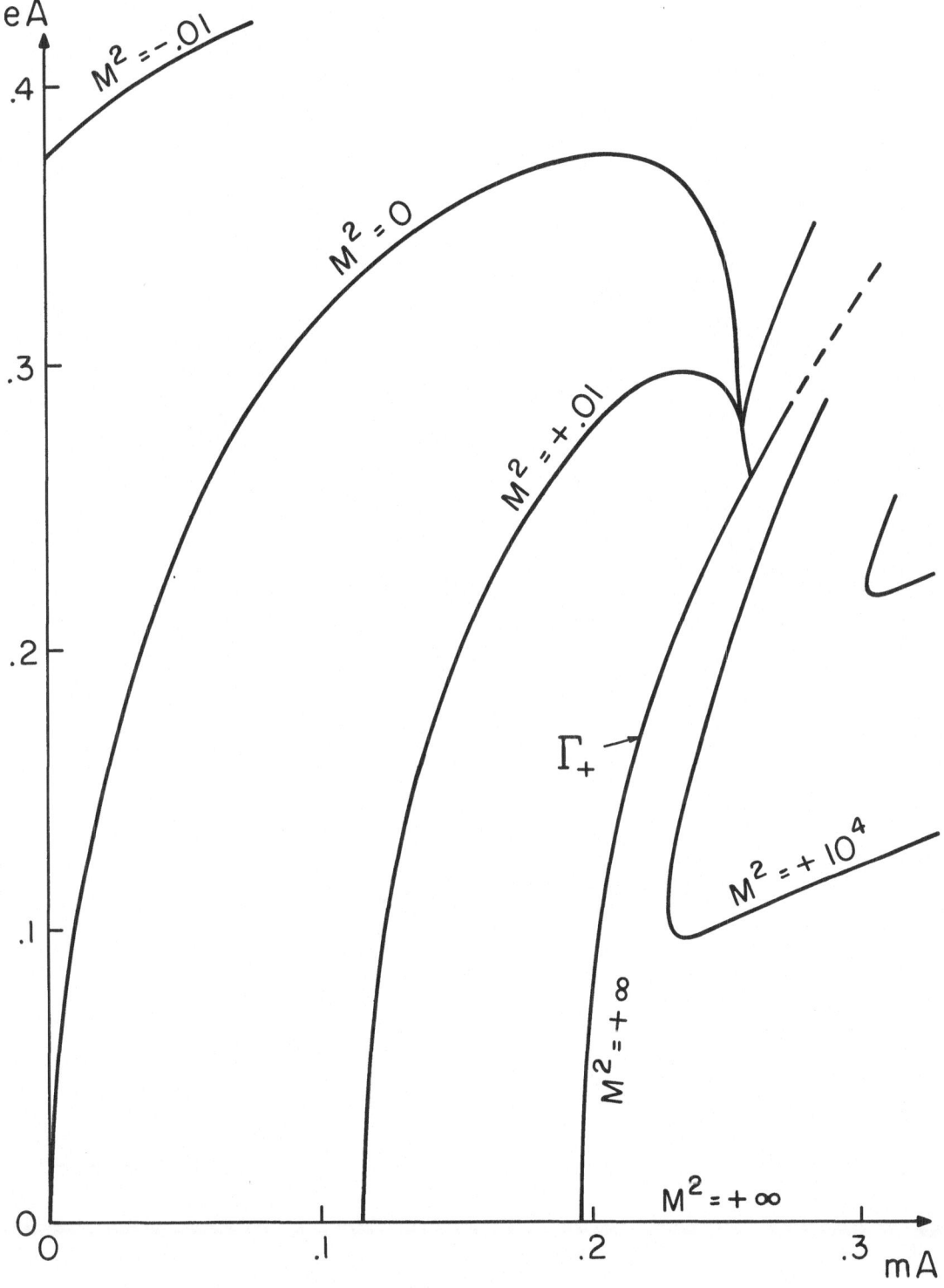

Figure 16

The Bondi rest mass M^2 evaluated at $\bar{u} = 0$.

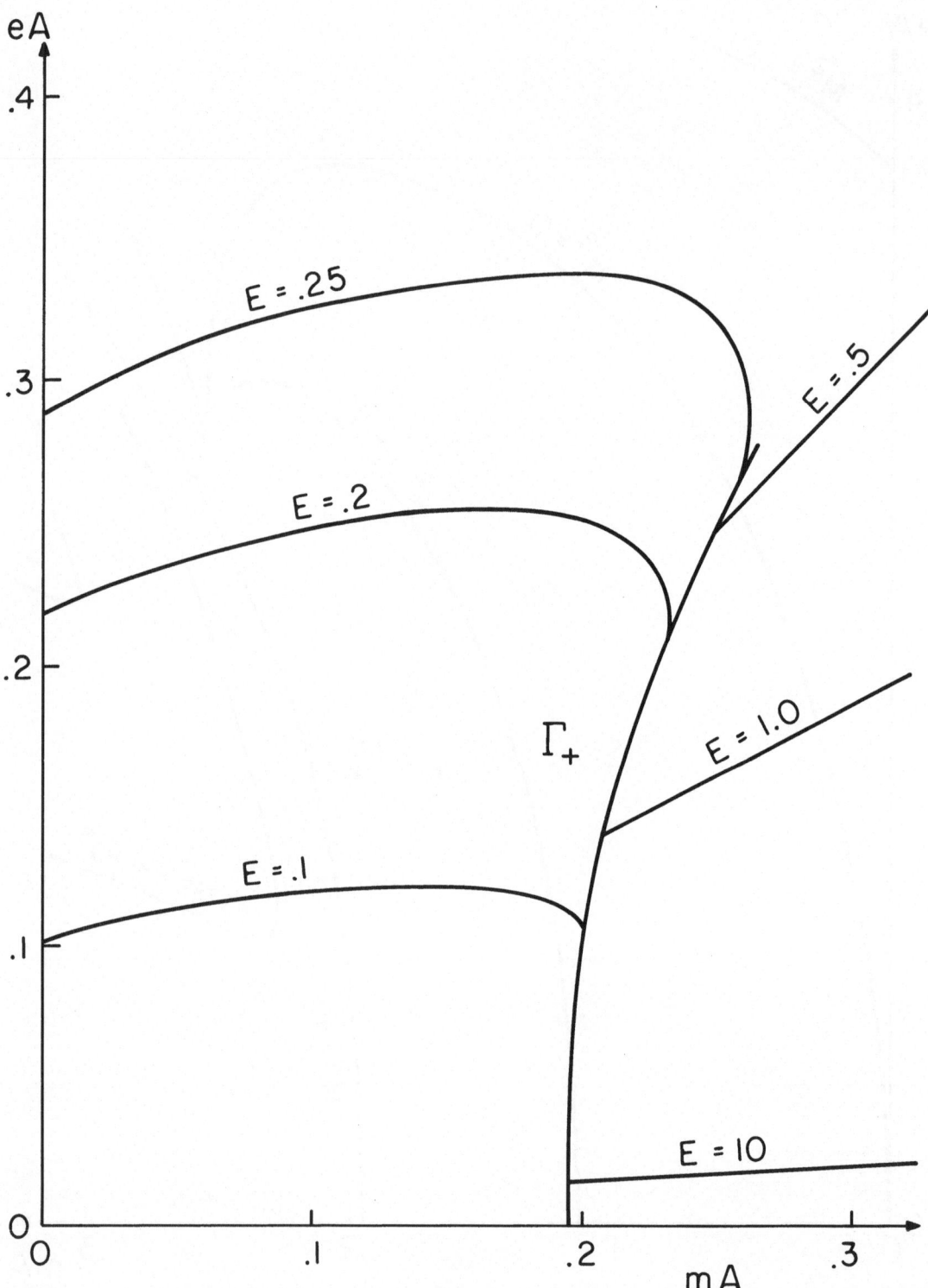

Figure 17

The Bondi charge E.

REFERENCES

1. Sometimes also called a Lorentz metric, or by relativists, simply a metric.

2. W. Kinnersley and M. Walker, Phys. Rev., to appear.

3. T. Levi-Civita, Atti del. Acad. dei Lincei:Rendiconti 27, 343 (1918).

4. W. Kinnersley, unpublished thesis, California Institute of Technology, 1969.

5. These generators are of course the flat space limits of the outgoing principal null congruence with tangent ℓ^a.

6. B. Carter, J. Math. Phys. 10, 70 (1969).

7. M. Walker, J. Math. Phys. 11, 2280 (1970).

8. E. Newman and T. Unti, J. Math. Phys. 3, 891(1962).

9. E. Newman, T. Unti, and L. Tamburino, J. Math. Phys. 4, 915(1963).

10. R. Sachs, Phys. Rev. 128, 2851(1962).

11. A. Janis and E. Newman, J. Math. Phys. 6, 902(1965).

12. R. Penrose, in Battelle Rencontres (ed. B. S. DeWitt and J. A. Wheeler; Benjamin, New York, 1968).

13. E. Newman and R. Penrose, Proc. Roy. Soc. (London) A305, 175(1968).

CONSERVATION LAWS ON MANIFOLDS

A. P. Stone

Department of Mathematics, University of Illinois at Chicago Circle [*]

I.   Introduction

Conservation laws arise in physics in several ways. We consider one of these possibilities and use it to motivate the definition of a conservation law on a manifold. Suppose that one has the following first order system of linear homogeneous partial differential equations in two independent variables,

$$(1) \quad \frac{\partial}{\partial t} U + (h) \frac{\partial}{\partial x} U = 0 ,$$

where U is a column vector of n dependent variables and (h) is an nxn matrix depending on U. If one can premultiply (1) by a suitable row vector so that the result has the form of a scalar equation

$$(2) \quad \frac{\partial f}{\partial t} + \frac{\partial g}{\partial x} = 0 ,$$

then (1) is said to contain the conservation law (2). More generally if one can premultiply (1) by a suitable non-singular matrix so that the result is

$$(3) \quad \frac{\partial V}{\partial t} + \frac{\partial W}{\partial x} = 0 ,$$

where V and W are column vectors depending on U, then (1) is said to contain the system of conservation laws (3). The possibility of finding such a non-singular matrix is of interest both for hyperbolic and elliptic partial differential equations. In the hyperbolic case one finds that the notion of a weak

*Present address: University of New Mexico, Albuquerque N. M.

solution of (1) depends on the possibility of expressing that system in terms
of conservation laws.

As an example, let us consider the case of one dimensional time-dependent,
non-isentropic, adiabatic fluid flow. The situation is described by the equations

$$(4) \quad \frac{\partial}{\partial t} \begin{pmatrix} \rho \\ u \\ p \end{pmatrix} + \begin{pmatrix} u & \rho & 0 \\ 0 & u & \frac{1}{\rho} \\ 0 & \gamma p & u \end{pmatrix} \frac{\partial}{\partial x} \begin{pmatrix} \rho \\ u \\ p \end{pmatrix} = \begin{pmatrix} 0 \\ 0 \\ 0 \end{pmatrix} \quad ,$$

where ρ, u, p, and γ are respectively the density, velocity, pressure, and
adiabatic constant. One also assumes $\gamma > 1$; if $\rho > 0$ we may then
premultiply (4) by the non-singular matrix

$$\begin{pmatrix} 1 & 0 & 0 \\ u & \rho & 0 \\ u^2 & 2\rho u & \frac{2}{\gamma-1} \end{pmatrix}$$

to obtain a system of conservation laws

$$\frac{\partial}{\partial t} \begin{pmatrix} \rho \\ \rho u \\ \rho u^2 + \frac{2p}{\gamma-1} \end{pmatrix} + \frac{\partial}{\partial x} \begin{pmatrix} \rho u \\ \rho u^2 + p \\ \rho u^3 + \frac{2\gamma}{\gamma-1} \rho u \end{pmatrix} = \begin{pmatrix} 0 \\ 0 \\ 0 \end{pmatrix} \quad ,$$

which expresses the conservation of mass, momentum, and energy.

We conclude with the remark that conservation laws may also arise from
variational principles, though this is not of interest here.

II. Conservation laws

The question of expressing a system (1) in the form (3) may be re-
formulated as a problem on manifolds. Specifically we let M and N represent
analytic manifolds with coordinates $(u^1,...,u^n)$ and (t,x) respectively, A
the ring of germs of analytic functions on M, E the localization of the module
of vector fields on M and $\mathcal{E}$ the dual module of differential forms on M.
Let $\underline{h}$ be an endomorphism of E; that is, $\underline{h} \in \text{Hom}_A(E,E)$. If $L \in E$ and $\theta \in \mathcal{E}$

we then describe the adjoint of $\underline{h}$ by the convention that $L\underline{h}(\theta) = L(\underline{h}\theta)$ and thus we do not distinguish notationally between $\underline{h}$ and its adjoint. Let

$$\frac{\partial}{\partial u^i} \underline{h} = h_i^\alpha \frac{\partial}{\partial u^\alpha}$$ and so with respect to a local basis $\left\{ \frac{\partial}{\partial u^i} \right\}_1^n$ of E, $\underline{h}$

is represented by the matrix (h_j^i). The problem associated with the system (1)

is to find a smooth map $N \xrightarrow{\;\Phi\;} M$ satisfying certain side conditions with a

differential $\dot{\Phi}$ such that $\frac{\partial}{\partial t}\dot{\Phi} + \frac{\partial}{\partial x}\dot{\Phi}\underline{h} = 0$ on M. Thus if $\underline{h}$ is replaced by

its adjoint then $(\frac{\partial}{\partial t}\dot{\Phi})(\theta) + (\frac{\partial}{\partial x}\dot{\Phi})(\underline{h}\theta) = 0$ for any $\theta \in \mathcal{E}$,

and moreover θ is a conservation law if and only if θ and $\underline{h}\theta$ are exact. Thus we

can eliminate the manifold N from consideration and define a conservation law

on M as follows:

Definition 1: $\theta \in \mathcal{E}$ is a <u>conservation law</u> for $\underline{h}$ if and only if θ and $\underline{h}\theta$

are exact.

The problem in general is that of finding all conservation laws for a given

$\underline{h}$. In the next section we give a solution to this problem in the case that $\underline{h}$

has distinct eigenvalues and vanishing Nijenhuis tensor.

III. The Nijenhuis tensor

An element $[\underline{h},\underline{h}]$ of $\mathrm{Hom}(\mathcal{E}, \mathcal{E} \wedge \mathcal{E})$ is defined in this section. It

is called the Nijenhuis torsion of $\underline{h}$ and its vanishing provides an integrability

condition which guarantees the existence of certain coordinates. The Nijenhuis

tensor plays a crucial role in the study of complex manifolds. Before its

definition is given, we note that $\underline{h}$ induces endomorphisms of $\mathcal{E} \wedge \mathcal{E}$ which are

defined by setting

$$h^{(1)}(\theta \wedge \phi) = \underline{h}\theta \wedge \phi + \theta \wedge \underline{h}\phi$$

$$h^{(2)}(\theta \wedge \phi) = \underline{h}\theta \wedge \underline{h}\phi$$

for any $\theta, \phi \in \mathcal{E}$. That these are well-defined is verified by considering the

appropriate endomorphisms of $\mathcal{E} \otimes \mathcal{E}$ and observing that they are stable on the

kernel of the natural map $\mathcal{E} \otimes \mathcal{E} \longrightarrow \mathcal{E} \wedge \mathcal{E}$. Thus if d denotes the exterior differentiation operator, then we define the Nijenhuis torsion of $\underline{h}$

by setting $[\underline{h},\underline{h}]\Theta = -h^{(2)}d\Theta + h^{(1)}d\underline{h}\Theta - d\underline{h}^2\Theta$ for any $\Theta \in \mathcal{E}$.

It is easy to verify that $[\underline{h},\underline{h}]$ is a homomorphism over A. The bracket $[\underline{h},\underline{h}]$ can also be regarded as an element of $\text{Hom}(E \wedge E, E)$ and this alternative definition is the one which appears in $[1]$.

If $\Theta \in \mathcal{E}$ is a conservation law for $\underline{h}$, then the condition $[\underline{h},\underline{h}] = 0$ implies $d\underline{h}^2\Theta = 0$. Hence since closed forms are locally exact, $\underline{h}\Theta$ is also a conservation law for $\underline{h}$ and one may then establish the following theorem.

<u>Theorem</u>: If $[\underline{h},\underline{h}] = 0$ and Θ is a conservation law for $\underline{h}$, then $\underline{h}^i\Theta$ is a conservation law for $\underline{h}$ for any positive integer i.

We remark that the preceding remark also holds globally if the first Betti number is zero.

If we now suppose that $\underline{h}$ has real-valued distinct eigenvalues $\lambda_1,\dots,\lambda_n$ with corresponding eigenvector fields, then the condition $[\underline{h},\underline{h}] = 0$ guarantees the existence of coordinates $x_1,\dots,x_n$ such that $X_i = \frac{\partial}{\partial x_i}$ and $\lambda_i = \lambda_i(x_i)$, $i = 1,\dots n$. This fact is established in $[3]$.

Thus one may take $\{dx_1,\dots,dx_n\}$ as an eigenform (local) basis for $\mathcal{E}$, and moreover this basis is a basis of conservation laws since $\underline{h}dx_i = \lambda_i(x_i)dx_i$ is exact.

We conclude this section with the observation that given any $\Theta \in \mathcal{E}$, then Θ is a conservation law for $\underline{h}$ if and only if $\Theta = \sum_{i=1}^{n} \alpha_i(x_i)dx_i$. It is assumed that $\underline{h}$ has distinct eigenvalues and vanishing Nijenhuis tensor.

IV. Other problems

The problem of conservation laws may be generalized in a number of directions. One possibility would be an obvious generalization of the system (1) to a system $\frac{\partial}{\partial t}U + (h)\frac{\partial}{\partial x}V + (k)\frac{\partial}{\partial y}W + (\underline{\ell})\frac{\partial}{\partial z}X = 0$. The corresponding problem for manifolds would then consist of finding conservation laws for given endomorphisms $\underline{h}$, $\underline{k}$, and $\underline{\ell}$; this problem is considered in $[5]$.

Another generalization is possible if the module $\mathcal{E}$ is replaced by $\Lambda^p \mathcal{E}$, the module of differential forms of degree p, and if $\underline{h} \in \mathrm{Hom}(\mathcal{E},\mathcal{E})$ is replaced by $T \in \mathrm{Hom}(\Lambda^p\mathcal{E}, \Lambda^p\mathcal{E})$. The question then becomes one of finding p-forms Ω such that Ω and $T\Omega$ are exact when T is given. This problem is introduced in [6].

Finally, the problem discussed in this paper may also be modified in a number of ways. For example, the problem treated here involved a real analytic manifold and all results were local in nature. Thus one might consider C^r manifolds and pose the problem of conservation laws as a global problem.

REFERENCES

1. A. Frolicher and A. Nijenhuis, Theory of vector valued differential forms, I. Proc. Ned. Akad. Wet. Amsterdam, 59, (1956), 338-359.

2. P. D. Lax, Hyperbolic systems of conservation laws, II. Comm. Pure and Applied Math. 10, (1957), 537-566.

3. A. Nijenhuis, X_{n-1} -forming sets of eigenvectors, Proc. Ned. Akad. Wet. Amsterdam 54, (1951), 200-212.

4. H. Osborn, Les lois de conservation, Ann. Inst. Fourier, Grenoble, 14, (1964), 71-82.

5. A. Stone, Generalized conservation laws, Proc. AMS. 18, (1967), 868-873.

6. A. Stone, Higher order conservation laws, J. Diff. Geometry, 3, (1969), 447-456.

STRUCTURE OF SINGULARITIES

M. Demianski[*]

University of Pittsburgh

The problem of singularities or rather singular spacetimes is the most
intriguing problem of present day general relativity. The occurrence of singularities is a well-known fact in classical field theories and usually it indicates
a limitation of the theory. We quite well understand now what the singularities in
hydrodynamics and electrodynamics are. We have learned how to describe them in
precise mathematical terms. A good example is provided by the shock waves in
hydrodynamics. The situation in general relativity is, however, much more
complicated. In any other field theory when we face the problem of singularities,
the spacetime which is its arena is given a priori; in general relativity we are
trying to describe singularities of a field which itself determines the properties
of spacetime. In this talk I will briefly review the definitions of singular
spacetimes and discuss various attempts to describe the structure of singularities.
At the moment we do not have a completely satisfactory definition of a singular
spacetime. All of the proposed definitions to date are related to the problem of
extensions and geodesic completeness. The main difficulty is that singular points
cannot be represented as points of the spacetime manifold; they appear as boundary
points. For a given manifold M, with the standard topological and differentiable
structure, there is no unique way to attach to it a set of boundary points. (The

*On leave of absence from University of Warsaw. Supported in part by the
National Science Foundation, Science Development Grant No. GU - 3184.

proposed prescriptions always assume additional structure and in general do not define a unique boundary.)

In order that a manifold M be an acceptable model of the Universe, we will require that M be inextendible, where by inextendible we mean that M is not isometric to a (proper) subset of some other spacetime M'. We would be tempted to call an inextendible spacetime M singular if:

a) at least one reasonable timelike world-line (say with a bounded acceleration) has a finite proper length,

b) certain scalar invariants of the Riemannian tensor blow up along certain timelike curves.

How to incorporate these conditions into a sensible, practical definition of a singular spacetime is a hard problem, some aspects of which are stated in appendix A.

We will call an inextendible spacetime M singular if it is <u>timelike geodesically incomplete</u>. (A spacetime is timelike geodesically incomplete if there exists a timelike geodesic in M which cannot be extended to an arbitrarily large value of the proper length). This definition is useful in proving theorems on the existence of singularities but it is not completely satisfactory since there is an example of a spacetime which is <u>timelike geodesically complete with a timelike line of bounded acceleration and finite proper length</u> [2].

A spacetime M (not necessarily inextendible) will be called singular if every extension M' of M, which is itself inextendible, is singular. With this definition of singular spacetime we have the following equivalence for inextendible spacetimes:

singular <=> timelike geodesically incomplete.

A spacetime which is singular according to this definition we will call <u>g-singular</u>. So a spacetime is g-singular if every extension of M is timelike geodesically incomplete. This means that only a geodesic incompleteness which cannot be removed by an extension counts as a singularity.

There is no natural way of obtaining from a singular spacetime a set of singular points whose local properties can be studied in detail. There do, however, exist a few procedures that one can follow in attempting to answer questions about the properties of singularities.

I. The fact that a spacetime is singular is indicated by the existence of in-
    extendible timelike geodesics. Therefore it is natural to use them in a
    process of defining a boundary of a singular spacetime which we will call
    the g-boundary. To define points of the g-boundary we proceed in the
    following way. We define equivalence relations among the timelike incomplete
    geodesics as follows. We will say that two geodesics γ_1 and γ_2 belong to the
    same equivalence class provided that γ_2 enters and remains in the region
    determined by all points which lie on geodesics obtained by infinitesimal
    variations of γ_1. An equivalence class of incomplete geodesics represents
    a point of the g-boundary. The g-boundary can be endowed with a topological,
    causal and in some cases metric   structure (See appendix B) [3,6].

II. This procedure can be modified and based on inextendible timelike curves
    (not necessarily geodesics). Consider the set of all inextendible timelike
    curves on a spacetime M. We will say that two inextendible timelike curves
    γ_1 and γ_2 are equivalent provided that the past of γ_1 is identical to the
    past of γ_2; this is an equivalence relation. With each of the equivalence
    classes of inextendible timelike curves we associate an "ideal point". The
    set of ideal points should be divided into classes of points representing
    true singular points and those representing points at infinity. This
    distinction is difficult to make, and no satisfactory method of distin-
    guishing between the two cases is known. The set of ideal points has very
    little structure. A reasonable topology has not been found on this set. The
    method of ideal points does not, in certain cases, differentiate between
    what one might consider to be different singular points [5].

III. Embedding theorems.

There exists a theorem (due to Clarke) that any spacetime can be isometrically embedded in a 90 dimensional flat space of signature (3,87). Because of
the high dimensionality of the space into which a spacetime may be embedded

it is practically hopeless to try to classify and describe properties of singular spacetimes using this technique [1].

A new approach to the problem of defining a singular spacetime and describing singular points was recently proposed by B. Schmidt [7]. This approach is based on the structure of the bundle of linear frames F(M). (See appendix C).

Let $p \in F(M)$, $p = (x, X_{(x)i})$ where $x \in M$ and $X_{(x)i} \in T_{(x)}(M)$. The projection $\pi: F(M) \rightarrow M$ is defined by $\pi(p) \rightarrow x$. The general linear group GL(n,R) acts on F(M) effectively (to the right); given $F(M) \ni p = (x, X_{(x)i})$, $a = (a_i^j) \in GL(n,R)$, then

$pa = (x, X_{(x)i} a_j^i)$; $p(ab) = (pa)b$.

We endow the bundle of linear frames F(M) with a linear connection Γ. The bundle of linear frames with a connection is parallelizable, so there exists a canonical set of linearly independent vectors which constitute a basis at each point of T(F(M)). The structure of the bundle of linear frames F(M) with a linear connection Γ also makes it possible to introduce into T(F(M)), in a natural way, a positive definite metric. Let $h_{ij}(i,j = 1,2,...,n)$ be a positive definite metric in R^n (for the time being let $h_{ij} = \text{Diag}(+,+,...,+)$), X_p and Y_p vectors tangent to F(M) at p; then we define a metric by

$$g(X,Y) = h_{ij} \theta^i(X) \theta^j(Y) + h_{ij} h^{k\ell} W_k^i(X) W_\ell^j(Y) ,$$

where h^{ij} is the inverse matrix of h_{ij}, and $\theta^i(X)$, $W_k^i(X)$ are defined in appendix C. This defines a positive definite metric on T(F(M)). With this choice of the metric the basis vectors of T(F(M)) are orthonormal.

The metric g(X,Y) enables us to introduce a distance function into a connected component $F_c(M)$ of F(M). In general, when M is orientable, F(M) consists of two connected components and when M is non-orientable F(M) itself is connected. Consider a connected component $F_c(M)$ of F(M). The distance function d(p,q) between p and $q \in F(M)$ is defined in the standard way as an upper lower bound of lengths of all piecewise differentiable curves joining p and q. With this distance function $F_c(M)$ becomes a metric space in the topological sense.

With a positive definite metric defined on $T(F(M))$ we can complete $F_c(M)$, obtaining $\widetilde{F_c(M)}$, where the points in $\widetilde{F_c(M)}$ represent equivalence classes of Cauchy sequences in $F_c(M)$. The $\widetilde{F_c(M)}$ obtained in this way is a unique, up to isomorphism, complete metric space. Of course, $F_c(M) \subset \widetilde{F_c(M)}$; the embedding is canonical.

Consider $\widetilde{F_c(M)}$ and the set of points constituting its boundary $\partial\widetilde{F_c(M)}$:

$$\partial\widetilde{F_c(M)} = \{p \; \epsilon \; \widetilde{F_c(M)} : p \notin F_c(M)\}.$$

The action of the general linear group can be easily extended to $\widetilde{F_c(M)}$. By $\widetilde{M}$ let us denote the set of orbits of $\widetilde{F_c(M)}$ under the action of the general linear group, and correspondingly by $\partial\widetilde{M}$ the set of orbits of $\partial\widetilde{F_c(M)}$. The projection operator $\Pi : F(M) \to M$ can be extended to $\Pi : \widetilde{F_c(M)} \to \widetilde{M}$ by defining $\Pi(p)$, where $p \; \epsilon \; \widetilde{F_c(M)}$, to be the orbit through p.

$\widetilde{F_c(M)}$ is endowed with a topology induced by the metric. We can make $\widetilde{M}$ into a topological space by requiring that $O\epsilon\widetilde{M}$ be open iff $\Pi^{-1}(0)$ is open in $\widetilde{F_c(M)}$. This is the weakest topology in which Π is continuous.

The prescription presented provides us with a method of attaching boundary points ∂M to M. The set of points ∂M we will call a b-boundary. It would be premature however to call all points of ∂M singular, because, for example, M could be extendible and the appearance of boundary points indicates that we are considering only part of a spacetime.

Following Schmidt we will define a singularity as a point of the b-boundary ∂M which is contained in the b-boundary of every extension of M. The definition as it stands is not too practical. To decide which points of $\partial\widetilde{M}$ represent singularities one must perform all possible extensions of M. It would be nice to have a criterion which would enable us to determine singular points of ∂M without actually performing these extensions.

A few questions arise naturally in connection with this definition. First, since the positive definite metric was introduced by choosing a particular basis

for R^n and for the Lie algebra $gl(n)$, does the b-boundary depend on this choice?
It is easy to see using the properties of the standard horizontal vector fields
$B(e_n)$ and the fundamental vector fields E_j^i that the distance function does not
depend on the choice of basis for R^n or for the Lie algebra of $gl(n)$. The b-
boundary $\partial\tilde{M}$ does not depend on the choice of the positive definite metric h_{ij}
either. This follows from the fact that given a linear connection of $F(M)$ the
distance functions are uniformly equivalent and as a matter of fact it is the
uniform structure on $F_c(M)$ which is uniquely defined by a linear connection in
$F(M)$. Therefore M and ∂M are independent of the particular choice of h_{ij}. (See
appendix F.)

Secondly, the previously proposed definition of a singular spacetime was
based on geodesic incompleteness. How are these definitions related?

A geodesic in M is represented in $F(M)$ by a horizontal line which is an
integral curve of a standard horizontal vector field. If $\tau^* = p_t$ with $t \in [0,1]$
is an integral curve of a standard horizontal vector field $B_{p_t}(\xi)$, then the
length of τ^* is given by:

$$L = \int_o^1 \left[h_{ab} \; \theta^a(B_{p_t}(\xi)) \; \theta^b(B_{p_t}(\xi)) \right]^{\frac{1}{2}} dt$$

The tangent vector to the geodesic $\tau = \pi(\tau^*) = X_t$ is

$$\dot{X}_t = \Pi_*(B_{p_t}(\xi)) = \hat{p}(\xi) = \xi^a B_{(X_t)_a} = \theta^a(B_{(p_t)}(\xi)) \quad B_{(X_t)_a} ,$$

(for an explanation see appendix C); hence $\theta^b(B_{p_t}(\xi))$
are the components of the tangent vector $\dot{X}_t$ of the geodesic
$X(t)$ in the frame $B_{(X_t)_a}$. We can look upon this result in the following way: the
length of a horizontal curve τ^* representing a geodesic τ in M is equal to the
length of a curve X_t in a Riemannian space with a positive definite metric h_{ab}
obtained by choosing an arbitrary point and a basis and then drawing a curve X_t
in such a way that the components of the tangent vector in the parallelly
propagated frame should be equal to $\theta^b(B_{p_t}(\xi))$.

Thus every incomplete geodesic in M determines a point of the b-boundary $\partial\tilde{M}$.

The class of curves which determine points of $\partial\tilde{M}$ is much larger, however. To describe this class of curves, consider an inextendible curve X_t in M and a frame parallelly propagated along this curve. Construct a curve $\tilde{X}_t$ in a Riemannian space R with a positive definite metric h_{ab} and a metric connection Γ, and choose a frame $\tilde{p}_t$ parallelly propagated along $\tilde{X}_t$ such that $\tilde{X}_t$ has the same components in the frame $\tilde{p}_t$ as X_t in the frame p_t. The inextendible curve X_t defines a point of $\partial\tilde{M}$ if and only if $\tilde{X}_t$ is extendible in R. It is not obvious that all points of $\partial\tilde{M}$ can be obtained in this way. That it is so is seen from the following theorem (Schmidt):

Theorem I: If V_n is a Cauchy sequence defining a point of $\partial\tilde{M}$, then there is a Cauchy sequence $\{u_n\}$ on a horizontal curve which determines the boundary point. (For a proof see appendix D.)

$\tilde{M}$ is endowed with a topology so that it is a topological space; we can say something more about its structure. The connected component of the bundle of linear frames $F_c(M)$ is a Riemannian space, hence $F_c(M)$ is paracompact and satisfies the second axiom of countability, being a metric space. Therefore $\widetilde{F_c(M)}$ is second countable and this is also true for $\tilde{M}$. $\widetilde{F_c(M)}$ is connected, therefore so is $\tilde{M}$. (For definitions see appendix E.)

$\tilde{M}$ as a topological space is not in general locally compact and is not necessarily Hausdorff. From the point of view of separation axioms the topological space $\tilde{M}$ is at most T^o as is indicated by the following theorem (Schmidt):

Theorem II: Suppose u_n is a Cauchy sequence in F(M) without limit in F(M), such that $\pi(u_n)$ is contained in a compact set of M. Then there exists xϵM such that $\pi^{-1}(x)$ is not complete and $\tilde{M}$ is at most T^o.

(For a proof see appendix D.)

If M is endowed with a chronological relationship, this relationship can be extended into $\tilde{M}$. M is a dense subset of $\tilde{M}$ (with canonical embedding), so to introduce a causal structure in $\tilde{M}$ it will be enough to define causal relations between any two points of $\partial\tilde{M}$ and points of M and $\partial\tilde{M}$.

Consider two points X and Z of $\partial\tilde{M}$. According to Theorem I there exist

Cauchy sequences $\{u_n\}$ and $\{v_n\}$ on horizontal curves p_t and g_t which determine

boundary points X and Z. By projecting p_t and g_t onto M we obtain curves X_t and

Z_t, and a sequence of points $\Pi(u_n)$ on X_t and $\Pi(v_n)$ on Z_t. We will say that X

<u>chronologically precedes</u> Z, and write $X \ll Z$ if there exist sequences $\{u_n\}$ and $\{v_n\}$

and numbers N_1 and N_2 such that for every $n > N_1$, $\Pi(v_n)$ is contained in $I^+ (\Pi(u_n)$

for every $m > N_2$.

If $x \in M$ and $Z \in \partial M$ the corresponding definition is more complicated. Let p_t

be a horizontal curve and $\{u_n\}$ a Cauchy sequence on p_t defining Z. We will say

that X <u>chronologically precedes</u> Z if there exists an open neighborhood 0 of x and

a number N such that $0 \subset I^-(\Pi(u_n))$ for all $n > N$.

Acknowledgement

I would like to thank Dr. A. Held for reading the manuscript and for many

comments and suggestions.

Appendix A

Definition of a Singular Spacetime.

In an attempt to define a singular spacetime one can draw an analogy with

other classical field theories, such as electrodynamics or hydrodynamics. In

electrodynamics, for example, we call a solution of Maxwell's equations singular

if the field is infinite and therefore undefined at a certain point of the

background Minkowski spacetime. By analogy then, one can say that the spacetime

is singular if the metric tensor blows up at some point. This can be due to the

choice of coordinate system, however, but even if not, we always can cut out from

the spacetime manifold a region where the metric tensor is undefined. From this

we conclude that a definition of a singular spacetime cannot be based on behavior

of components of tensor fields and also that the singular points cannot be represented as points of a spacetime manifold.

Because it is always possible to cut out a region from spacetime we would like to be sure that no region has been removed from any spacetime which we would like to call nonsingular. The question whether some regions of spacetime were removed or not can be investigated by studying the proper distance along geodesics in the spacetime. The spacetime will be called geodesically incomplete if there exists a geodesic which cannot be extended to arbitrarily large values of an affine parameter.

In spacetime we have three different types of geodesics; null, timelike and spacelike. As there are examples of spacetimes which are timelike complete, and null and spacelike incomplete, one can have one of the five following possibilities:

A spacetime can be:

complete	incomplete
timelike	spacelike and null
spacelike	timelike and null
null	timelike and spacelike
timelike and null	spacelike
spacelike and null	timelike

From the physical point of view we would like to call a spacetime M singular if it is timelike or null geodesically incomplete. The incompleteness can indicate that the spacetime is truly singular or that we are considering only an extendible spacetime manifold. The second possibility can be removed by first performing an extension of M to an inextendible spacetime M^1 and thereafter investigating whether M^1 is timelike geodesically complete or not.

So finally, we will call a spacetime M singular if every extension of M which is itself inextendible, is timelike geodesically incomplete [2, 4, 6].

This definition is not satisfactory as there exists an example, due to Geroch, of a spacetime which is timelike geodesically complete but contains a timelike line with bounded acceleration which has a finite proper length [2].

Appendix B

Definition and Structure of the g-boundary

The structure of the g-boundary was extensively studied by Geroch [3] and we will follow his approach. A geodesic in a spacetime M is uniquely defined if one specifies a point $p \in M$ and a nonzero vector $X_p \in T_p(M)$. The collection of pairs (p, X_p) we will denote by G. G is the tangent bundle of M with the zero vector omitted. Each element of G, (p, X_p) uniquely determines that geodesic in M which begins at p and has tangent vector at p equal to X_p. Therefore there is a one to one correspondence between elements of G and geodesics in M.

By a geodesic we will mean here a curve which:

1) has one and point and has been extended as far as possible in some direction from that end point;

2) is a geodesic with an affine parametrization. The affine parameter vanishes at the end point and is positive elsewhere on the curve.

Define a scalar field ϕ on the manifold G as the total affine length of the corresponding geodesic in M. ϕ is infinite if and only if the geodesic is complete. Denote by G_I that subset of G on which ϕ is finite.

Define the manifold $H = G \times (0, \infty)$, and the following two subsets of H:

$$H_+ \equiv \{(p, X_p, a) \in H : \phi(p, X_p) > a\} \,,$$

$$H_o = \{(p, X_p, a) \in H : \phi(p, X_p) = a\} \,.$$

There is a natural map $\psi : H_+ \rightarrow M$ defined as follows: Given a point (p, X_p, a) of H_+, let $\psi(p, X_p, a)$ be that point of M which results from traversing an affine distance a along the geodesic (p, X_p).

Let 0 be any open set of M. We associate with 0 a subset $S(0)$ of G_I as follows:

$$S(0) \equiv \{(p, X_p) \in G_I : \text{ there exists an open set U in H containing the point } (p, X_p, \phi(p, X_p)) \text{ of } H_o \text{ such that } \psi(U \cap H_+) \subset 0\}.$$

One can easily verify that, given any two open sets 0_1 and 0_2 of M the subsets $S(0_1)$ and $S(0_2)$ obey

$$S(O_1) \cap S(O_2) = S(O_1 \cap O_2).$$

The collection of sets $S(O)$, when O ranges over all open sets of M, serves as a basis for the open sets of a topology on G_I. We use this topology on G_I to form equivalence classes of the elements of G_I as follows: α and β are two elements of G_I with $\alpha \approx \beta$ if every open set containing β also contains α. The relation $\approx$ is an equivalence relation. The collection of equivalence classes will be denoted by ∂ and called the g-boundary. The topology we have defined on G_I induces a topology on ∂.

So far the g-boundary exists only as an abstract topological space. We now attach ∂ to the spacetime M. Define

$$\overline{M} = M \cup \partial \qquad \text{(the disjoint union)} .$$

A subset (O, U) of $\overline{M}$, where O is an open set of M and U is an open set of ∂, will be called open in $\overline{M}$ if $S(O) \supset U$. One can verify that the intersection of two such open sets is open. These open sets are a basis for a topology on $\overline{M}$. We will call $\overline{M}$ the spacetime with g-boundary.

Causal Structure

Let C be any directed curve in M. We say C has an end point at the point $e \in \partial$ if for every open neighborhood (O,U) of e in $\overline{M}$, C enters and remains in O. A curve in M having at least one end point in each direction is said to connect these points. Let e be any point of ∂; we define the future of e:

$$I^+(e) = \{Q \in \overline{M} : \text{there exists a timelike curve } C \text{ in } M \text{ with a future} \\ \text{end point } Q \text{ and a past end point } e\}.$$

The past, $I^-(e)$, is similarly defined. $I^+(e)$ and $I^-(e)$ are well-defined for every g-boundary point e without any other restriction on the spacetime.

One can proceed further and define what one means by a spacelike or timelike g-boundary. These properties should be local properties, i.e., properties not affected by changes in the geometry of regions away from the g-boundary. We must define, therefore, the local future of a point of the g-boundary. Let $e \in \partial$, and let (0,U) be an open neighborhood of e in $\overline{M}$ such that U = S(0),   then:

I^+(e;0,U) = {the future of e in (0,U)}, and similarly for the local past,

I^-(e;0,U). We say that ∂ is spacelike at e if there exists an open neighborhood

(0,U) such that U = S(0) of e in M and such that, for all e' ε U, there is an

open neighborhood (0',U'), U' = S(0'), of e' with

$$\left[I^+(e;0,U) \cup I^-(e;0,U) \right] \cap (0',U') = \phi .$$

We say that ∂ is timelike at e if for every open neighborhood (0,U) such that

S(0) = U of e in $\overline{M}$ there exist two points e' and e'' of U such that

I^+(e';0,U) $\cap$ I^-(e'';0,U) contains an open neighborhood of e in $\overline{M}$.

Differentiable Structure

Let 0 be any open set of the spacetime M, and let (0,U) be the corresponding

open set in $\overline{M}$ such that S(0) = U. Since the points of ∂ are equivalence classes

of elements of G_I, we have a natural map $\Pi: G_I \rightarrow \partial$. Define for each open set

0 of M, the following subset of G_I:

$\overline{J}(0)$ = {(p,X_p) ε Π^{-1}(U) : ϕ(p,X_p) = 1, and the geodesic in M associated
        with the element (p,X_p) of G_I lies entirely within the open set 0}.

We say that ∂ has a differentiable structure at the point e of ∂ if there exists

an open neighborhood (0,U) (S(0) = U) of e in $\overline{M}$ and a subset J of $\overline{J}(0)$ such that

the following conditions are satisfied:

1) J includes almost all points of $\overline{J}(0)$.

2) J is a differentiable submanifold of G.

3) For each element e of U, $A_e \equiv \Pi^{-1}$(e) $\cap$ J is a differentiable submanifold
    of J.

4) J may be written diffeomorphically as a cross product, A x B, of two
    differentiable manifolds, where for each element of B the submanifold
    A is one of the A_e. We have, therefore, a one-to-one mapping $\Lambda: U \rightarrow B$.
    We require further that this mapping be a homeomorphism.

The differentiable structure of B and the mapping $\Lambda: U \rightarrow B$ automatically

gives us a differentiable structure on the open region U of ∂.

From the definition we see that the differentiable structure, when defined, is unique. Note also that if a differentiable structure exists at a point e ε ∂, then the structure is defined in a neighborhood U of e. The collection of all points of ∂ at which a differentiable structure is defined forms a differentiable manifold.

Appendix C

Bundle of Linear Frames [8]

Consider an n-dimensional manifold M and let $T_x(M)$ denote the tangent space to M at x ε M, and let $X_{(x)i}$ ε $T_x(M)$, i = 1,2,...,n, be a set of linearly independent vectors which form a basis for $T_x(M)$. By a frame at a point x ε M we will mean p = (x, $X_{(x)i}$), (a point x of M and a set of n linearly independent vectors $X_{(x)i}$ ε $T_x(M)$). The collection of all frames we will denote by F(M).

We define the projection Π: F(M) → M by Π(x, $X_{(x)i}$) → x. The set $\Pi^{-1}(x)$ we will call a fibre. The general linear group GL(n,R) acts on F(M) (to the right) by p = (x, $X_{(x)i}$)

$$pa = (x, X_{(x)i} \, a_j^i), \text{ where } a = (a_j^i) \text{ ε GL(n,R), p(ab) = (pa)b.}$$

This action is effective since if pa = p for every p ε F(M) then a = id. With this structure it is easy to see how to make F(M) into a principal bundle which we shall call the bundle of linear frames.

Fundamental Vector Fields

Let A be an element of the Lie algebra $\mathfrak{gl}(n,R)$ of the general linear group GL(n,R) which acts on F(M). Exp (tA) is then a one-parameter group of transformations on F(M). Consider a point p ε F(M) and the trajectory of this point, p exp (tA), under the action of this group. The tangent vector to this trajectory at the point p

$$\tilde{A}_p = \frac{d}{dt} \, p \, \exp \, (tA) \, \Big|_{t=0}$$

we will call the fundamental vector field associated with A. The fundamental vector fields are tangent to the fibres at each point $p \in F(M)$; they span a subspace which we will call a vertical subspace and denote by V_p.

Canonical Form θ

A frame p gives an identification, $T_{\pi(p)}(M) \to R^n$ by

$$\hat{p} : R^n \to T_{\pi(p)}(M); \quad R^n \ni \xi = \xi^i e_i \to \xi^i X_{(\pi(p))i} \in T_{\pi(p)}(M) ,$$

where e_i is the standard basis for R^n. p is a vector space isomorphism. Using p we can define a canonical R^n-valued form on $F(M)$:

$$\theta(X)_p = p^{-1}(\Pi_* X_p); \quad X_p \in T_p(F(M)) .$$

θ may be interpreted as the lift to $F(M)$ of the coframes at $\Pi(p)$.

Linear Connection Γ on the Bundle of the Linear Frames

A linear connection Γ is an assignment of a subspace H_p of $T_p(F(M))$ to each $p \in F(M)$ such that:

1) $T_p(F(M)) = H_p \oplus V_p$ (direct sum).

2) $H_{pa} = R_{a*} H_p$, $p \in F(M)$, $a \in GL(n)$, and R_{a*} is the transformation of $T(F(M))$ induced by $R_a p = pa$.

3) H_p depends differentially on p.

Given a connection in $F(M)$ we define a 1-form ω on $F(M)$ with values in the Lie algebra $\mathfrak{gl}(n,R)$ of $GL(n,R)$. Let $X \in T_p(F(M))$; then $\omega(X) = A \in \mathfrak{gl}(n,R)$ is an element of the Lie algebra such that the corresponding fundamental vector field is equal to the vertical component of X:

$$\tilde{A}_p = \text{ver } X_p .$$

ω satisfies:

1) $\omega(\tilde{A}) = A$.

2) $(Ra)^* \omega = ad(a^{-1}) \omega$; $\omega(R_{a*}X) = ad(a^{-1}) \cdot \omega(X)$ for every $a \in GL(n,R)$ and every $X \in T(F(M))$.

Of course, $\omega(X_p) = 0 \Rightarrow X_p$ is horizontal.

Standard Horizontal Vector Fields

With each $\xi \in R^n$ we associate a standard horizontal vector field $B(\xi)$ on $F(M)$. For any $p \in F(M)$, $B_p(\xi)$ in the unique horizontal vector field at p such that:

$$\Pi_*(B_p(\xi)) = \hat{p}(\xi).$$

The standard horizontal vector field posesses the following properties:

1) $\theta(B(\xi)) = \xi$ when $\xi \in R^n$ and θ is the canonical form.
2) $R_{a*} B(\xi) = B(a^{-1}\xi)$, $a \in GL(n,R)$, $\xi \in R^n$.
3) If $\xi \neq 0$, $B(\xi)$ never vanishes.

Parallelization of $T(F(M))$

Let e_a, $a = 1,2,\dots,n$, be a natural basis of R^n and E_j^i any basis of $\mathfrak{gl}(n)$, so that:

$$\theta = \theta^a e_a , \qquad \omega = \omega_j^i E_i^j , \qquad a,i,j = 1,2,\dots,n .$$

The standard horizontal vector field corresponding to e_a we will denote by B_a and the fundamental vector fields corresponding to E_j^i by $\widetilde{E}_j^i$ respectively. The vectors B_a and $\widetilde{E}_j^i$ span the tangent space to $F(M)$ and the forms θ^a and ω_j^i span the cotangent space to $F(M)$. They are dual, as is seen from the relations:

$$\theta^a(B_b) = \delta_b^a , \qquad \theta^a(E_j^i) = 0 ,$$

$$\omega_j^i(B_a) = 0 , \qquad \omega_j^i(E_i^k) = \delta_i^i \delta_j^k .$$

The $n^2 + n$ vector fields $\{B_a, E_j^i\}$ define an absolute parallelism in $T(F(M))$; that is, they form a basis of $T_p(F(M))$ for every $p \in F(M)$.

Lift of a Vector Field

Given a vector field X on M, we can define a unique vector field on $F(M)$ by lifting X from $T(M)$ to $T(F(M))$; more precisely, if $X \in T_x(M)$, $p \in F(M)$ such that $\pi(p) = x$ then the unique horizontal vector field X^* ($\omega(X^*) = 0$) $X^* \in T_p(F(M))$ such that $\Pi(X^*) = X$ we will call the horizontal lift of X.

Lift of a Curve

Similarly we can define a horizontal lift of a curve in M. Let $J = X_t$ ($a < t \leq b$) be a (at least C') curve in M. A curve in F(M), $J^* = p_t$, such that $\Pi(J^*) = J$ and $\omega(p_t) = 0$ ($a \leq t \leq b$) we will call a horizontal lift of J. If we choose an arbitrary point $p \in F(M)$ such that $\Pi(p) = X_a$ then there always exists a unique lift J^* of J which starts from p.

If X^* is the vector field on F(M) which is a lift of $X \in T(M)$ then the integral curve of X^* through a point $p \in F(M)$ is a lift of the integral curve of X through the point $x = \Pi(p) \in M$.

Parallel Displacement, Geodesics

Let $J = x_t$, $0 \leq t \leq 1$, be a curve in M, and $p_0 \in F(M)$ such that $\Pi(p_0) = x_0$. The unique horizontal lift J^* of J through p_0 has the end point p_1: $\Pi(p_1) = x_1$. By varying p_0 in the fiber $\Pi^{-1}(X_0)$ we obtain a mapping of the fiber $\Pi^{-1}(x_0) \to \Pi^{-1}(x_1)$ which maps p_0 p_1. We will denote this mapping by J and call it __parallel displacement along__ J. J is an isomorphism of $\Pi^{-1}(p_0)$ onto $\Pi^{-1}(p_1)$. This follows from the fact that the parallel displacement along any curve J commutes with the action of GL(n) on F(M); $JoR_a = R_aoJ$ for every $a \in GL(n,R)$.

A curve on which the tangent vector is parallelly propagated is a geodesic. The projection onto M of any integral curve of a standard horizontal vector field on F(M) is a geodesic and conversely every geodesic lifted to F(M) is a integral curve of a standard horizontal vector field.

Completeness of a Linear Connection

We will say that a linear connection Γ of F(M) is complete if every geodesic can be extended to a geodesic $J = x_t$ with $-\infty \leq t \leq +\infty$ when t is an affine parameter. This can be stated equivalently in terms of standard horizontal vector fields. A linear connection Γ is complete if every standard horizontal vector field on F(M) is complete.

Appendix D

Proof of Theorem I:

There exists a curve $p(t)$ of finite length such that $t \in [0,1)$ and a set of points along the curve $V_n = p(t_n)$ with $\lim_{n \to \infty} t_n = 1$. There is no t_N with the property that $p(t_n)$ is contained in one fibre for $t_n > t_N$, because that would imply that V_n has a limit in this fibre. However, as the fibres are complete relative to the induced metric (every fibre is a homogeneous space), this is a contradiction. Hence $X_t = \Pi(p_t)$ is an inextendible curve in M.

Let u_0 be a frame at $X(0)$ and $u(t)$ the horizontal curve with $u(0) = u_0$ and $\Pi(u(t)) = X(t)$. Choose a frame $\tilde{u}_0$ at the origin of R^n endowed with the trivial connection and construct a curve $\tilde{X}(t)$ in R^n whose tangent vector has the same components in a frame parallel to $\tilde{u}_0$ as $X(t)$ in the frame $u(t)$. Consider the following mapping

$$\Phi : F(M) \ni \Pi^{-1}(X(t)) \to \Pi^{-1}(\tilde{x}(t)) \in F(R^n) \text{ defined by}$$

$$\Phi(u(t)) = \tilde{u}(t),$$

$$\Phi(R_a u(t)) = \tilde{R}_a \tilde{u}(t),$$

where $\tilde{u}(t)$ is the lift of $\tilde{X}(t)$ through u_0 and $\tilde{R}_a$ denotes the action of $GL(n,R)$ on the bundle of frames over R^n.

Φ maps the vector fields $\tilde{E}^i_j$ on $\Pi^{-1}(x(t))$ onto the vector fields $\tilde{\tilde{E}}^i_j$ on $\Pi^{-1}(\tilde{X}(t))$ and the curve $u(t)$ is mapped isometrically onto $\tilde{u}(t)$. This shows that Φ is an isometry of the Riemannian metric induced by the bundle metric on $\Pi^{-1}(x(t))$ and $\Pi^{-1}(\tilde{x}(t))$ as R_{a*} and $\tilde{R}_{a*}$ and in the same way on $\dot{u}(t)$ and $\dot{\tilde{u}}(t)$. More precisely, if $\dot{u}(t) = B(\xi(t))$, then $\dot{\tilde{u}}(t) = \tilde{B}(\xi(t))$ by the definition of u. The transformation properties of $B(\xi)$ and $\tilde{B}(\xi)$ imply $R_{a*} \dot{u}(t) = B(a^{-1}\xi(t))$ and $R_{a*} \dot{\tilde{u}}(t) = \tilde{B}(a^{-1}\xi(t))$. Hence the vector field $B(a^{-1}\xi(t))$ on $\Pi^{-1}(x(t))$ is mapped by Φ isometrically onto $\tilde{B}(a^{-1}\xi(t))$ on $\tilde{\Pi}^{-1}(\tilde{x}(t))$.

By the definition of X(t) the curve V(t) is contained in $\pi^{-1}(X(t))$ and has finite length. Therefore $\mathbf{\Phi}(V(t)) = V(t)$ has finite length in the trivial bundle over R^n. The bundle metric of this bundle is complete because it is a homogeneous Riemannian space. Hence X(t) is extendible and therefore u(t) has finite length. Via the isometry $\mathbf{\Phi}$ one finds that u(t) has finite length. Thus $u(t_n)$ is a Cauchy sequence in F(M) on a horizontal curve defining the same point of the b-boundary as V_n. Q.E.D.

Proof of Theorem II:

As $\pi(u_n)$ is contained in a compact set of M there is an infinite subsequence u'_n with $\lim \pi(u'_n) = x \in M$. Suppose $\pi^{-1}(X)$ is complete, hence closed in F(M); then $d(u'_n, \pi^{-1}(x))$ is always positive and there exist points $V_n \in \pi^{-1}(x)$ with $d(u'_n, \pi^{-1}(x)) = d(u'_n, V_n)$. But $\lim \pi(u'_n) = X$ implies $\lim d(u'_n, V_n) = 0$. Hence V_n is a Cauchy sequence with the same limit in $\partial \widetilde{F(M)}$ as u_n and therefore $\pi^{-1}(x)$ is not complete.

If $\pi^{-1}(x)$ is an incomplete fibre and V a limit of a Cauchy sequence in $\pi^{-1}(x)$ without limit in $\pi^{-1}(x)$, every open set containing V contains points of $\pi^{-1}(x)$. Thus every open set in M containing $\pi(V)$ contains x hence $\widetilde{M}$ is at most T^0. Q.E.D.

Appendix E

Definitions

Connected Component

A connected component E of a topological space X is a maximal connected subset of X.

Arcwise Connected

A space X is said to be arcwise connected if, given any two points a, b of X, there exists a continuous mapping of $[0,1]$ into X such that $a \to 0, b \to 1$.

Locally Compact

A topological space X is said to be locally compact if and only if each point of X has at least one compact neighborhood in X.

Paracompact

A covering A_i of a topological space X is said to be locally finite if for each $x \in X$ there exists a neighborhood U of X that intersects A_i for at most a finite set of indices i (this finite set depending in general on x and on U).

A covering B_j will be called a refinement of a covering A_i if and only if each B_j is a subset of some A_i.

A topological space is said to be paracompact if and only if it is regular and each open covering of X has an open locally finite refinement.

Separation Axioms

A topological space X is said to be:

1) T_0 if for any two distinct points x and y of X there exists either a neighborhood of x not containing y or a neighborhood of y not containing x.

2) T_1 if for any two distinct points x and y of X there exists a neighborhood of x not containing y and a neighborhood of y not containing x.

3) T_2 (or Hausdorff) if for any two distinct points x and y of X there exists a neighborhood of x and a neighborhood of y that are disjoint.

4) Regular if for each point x of X the closed neighborhoods of x form a neighborhood base at x.

5) T_3 if it is regular and T_1.

6) Normal if any two disjoint closed sets possess disjoint neighborhoods.

7) T_4 if it is normal and T_1.

Second Countable Spaces

A topological space (X, τ) is called a second countable space if there exists a countable base B for the topology τ.

Appendix F

Uniform Structures on a Set [9]

Let X be a set and $\Delta = \Delta(X) \subset X \times X$, the diagonal on X. By a uniform structure on X is meant a set $\mathcal{X}$ of subsets of $X \times X$ satisfying the following conditions:

a) $U \supset \Delta$ for each $U \in \mathcal{X}$.

b) If $U \in \mathcal{X}$, then $U^{-1} \in \mathcal{X}$.

c) If $U \in \mathcal{X}$, then there exists $U' \in \mathcal{X}$ such that $U' \circ U' \subset U$.

d) The intersection of two members of $\mathcal{X}$ also belongs to $\mathcal{X}$.

e) Any subset of $X \times X$, which contains a member of $\mathcal{X}$, itself belongs to $\mathcal{X}$.

By a uniform space is meant a pair $(X, \mathcal{X})$ comprising a set X and a uniform structure $\mathcal{X}$ on X.

If $(X, \mathcal{X})$ and $(Y, \mathcal{Y})$ are uniform spaces, and f a function on X onto Y, then f is said to be uniformly continuous if for each $V \subset Y \times Y$, the set

$$\{(X, X') \; \varepsilon \; X \times X \; : \; (f(x), f(x')) \; \varepsilon \; V\}$$

belongs to $\mathcal{X}$. f is then continuous on X into Y for the associated topologies.

References

1. Clarke, C. J. S., On the Global Isometric Embedding of Pseudo-Riemannian Manifolds, (preprint).

2. Geroch, R., What is a Singularity in General Relativity, Am. Phys. 48, 526, (1968).

3. Geroch, R., Local Characterization of Singularities in General Relativity (J. Math. Phys.) 9, 450, (1968).

4. Geroch, R., <u>Singularities</u>, Proceedings of the Cincinnati Conference on General Relativity.

5. Geroch, R., Kronheimer, F. H., Penrose, R., <u>Ideal Points in Spacetime</u>, Proc. Roy. Soc. (to appear)

6. Hawking, S. W., <u>Singularities and the Geometry of Spacetime</u>, Adams Prize Essay, 1966.

7. Schmidt, B. G., <u>A New Definition of Singularities in General Relativity</u>, (Preprint).

8. Kobayeshi, S., Nomizu, K., <u>Foundations of Differential Geometry</u> (Interscience, 1963).

9. Kelley, I. L., <u>General Topology</u> (D. van Nostrand Co. Inc., 1957).

LATTICE TRANSFORMATIONS AND CHARGE QUANTIZATION

Mayer Humi

Department of Mathematics
The University of Toronto

I. Introduction

There exist in physics two fundamental charges which give rise to long range forces. The first is the mass and the second is the electric charge. One can try to explain the existence of these charges in two ways. The first is a cosmological approach on the lines of Mach's principle and the second is the local point of view which is related to the local symmetries of the physical world. It is in this latter approach that the mass has its interpretation as a Casimir operator of the space-time symmetry group, i.e., the Poincaré group. However, no corresponding relationship is known to exist for the electric charge and it is our main objective in this paper to show the possible close connection between the electric charge and the dilatation operator on space-time.

The dilatation operator has played a restricted role in physics up to now [1,2] because it does not commute with the mass operator. We shall show, however, that if we break this symmetry in a special way then we can overcome this difficulty and then the dilatation operator displays the basic characteristics of the electric charge.

In section II we present the mathematical techniques which underlie the symmetry breaking; in section III we present the proposed identification, and we end in section IV with a discussion on the results.

II. Lattice Transformations

To start with let us consider the Weyl algebra in n-dimensions generated by P_i, Q_i $(i = 1,...,n)$ whose commutation relations (CR) are given by

$$[P_i, Q_j] = \delta_{ij} \qquad i, j = 1,...,n. \qquad (2\cdot1)$$

However, within the universal covering group of this algebra a group theoretical form of these (CR) has been developed [3,4] which is

$$e^{isP_i} e^{itQ_i} = e^{ist} e^{itQ_i} e^{isP_i}. \qquad (2\cdot2)$$

It has been noticed by mathematicians [3] and solid state physicists [4] that although these two forms are equivalent in many respects, equation $(2\cdot2)$ contains an extra and crucial piece of information which is not contained in equation $(2\cdot1)$. Thus it was realized that although the transformations e^{isP_i}, e^{itQi} do not commute in general, equation $(2\cdot2)$ implies that on a lattice of points which satisfy $st = 2\pi k$ they do commute.

However, it can be readily seen now that this procedure is extendible to other types of Lie algebra commutators. As a first example let us consider complex transformation groups for which two of the corresponding Lie algebra generators satisfy

$$[A,B] = iB \qquad (2\cdot3)$$

(or more generally $[A,B] = cB$ where c is some constant). In fact these (CR) would imply for the universal covering group of the algebra

$$e^{i\alpha A} e^{i\beta B} = \left(e^{i\beta B}\right)^{\exp(-\alpha)} \cdot e^{i\alpha A}. \qquad (2\cdot4)$$

Thus although the two transformations do not commute in general they <u>do commute</u> when α is restricted to the lattice $\alpha = 2\pi in$ where n is an integer.

As a second, more sophisticated, example of this technique, let us consider the following situation for the (CR) of the operators H, A, B, in a Lie algebra L :

$$[A,B] = H , \quad [H,A] = [H,B] = 0 . \qquad (2\cdot5)$$

It follows then that

$$e^{i\alpha A} \; e^{i\beta B} = e^{i\alpha\beta H} \; e^{i\beta B} \; e^{i\alpha A} \qquad (2\cdot6)$$

and therefore if we deal with a representation in which H is diagonal, i.e.,

$$H|f\rangle = m|f\rangle , \qquad (2\cdot7)$$

it follows that

$$e^{i\alpha A} \; e^{i\beta B} \; | f\rangle = e^{i\alpha\beta\, m} \; e^{i\beta B} \; e^{i\alpha A} \; | f\rangle \qquad (2\cdot8)$$

implies a local commutativity of $e^{i\alpha A}$, $e^{i\beta B}$ under suitable restrictions of α and β .

Let us remark that the above treatment was given in terms of abstract Lie groups and algebras. For certain realizations of algebras as differential operators it might happen that one or more operators become equal to unity on the above discussed lattice. In this case another slightly different realization can be found which resolves the degeneracy (as is shown in the next section) [5] .

Let us remark finally that equations $(2\cdot2)$, $(2\cdot8)$ require lattice restrictions of both operators for their commutativity while equation $(2\cdot4)$ requires lattice restrictions for one operator only. This fact will have a bearing on the discussion in the following sections.

III. Complex Dilatations and the Electric Charge

To apply the ideas of the last section let us consider the complex Poincare group [6], i.e., we allow the group parameter to assume complex values and adjoin to the algebra of this group the dilatation operator which satisfies the following (CR) with those of the Poincare algebra

$$\left[D, P_\mu\right] = iP_\mu \; , \qquad \left[D, M_{\mu\nu}\right] = 0 \; . \qquad (3\cdot1)$$

(We denote this algebra by g and the corresponding group G.)

A direct application of equation $(2\cdot4)$ now shows that $e^{i\alpha D}$ and $e^{i\beta P_\mu}$ commute when α is restricted to the lattice $\alpha = 2\pi i n$ (notice that β remains arbitrary and therefore space-time is still continuous). We conclude therefore that this breaking of the dilatation "symmetry" enables us to apply the group G to particles with non-zero mass.

To understand better the meaning of this result we note that the usual realization of g in four dimensions is

$$J_{\mu\nu} = i(X_\mu \frac{\partial}{\partial X_\nu} - X_\nu \frac{\partial}{\partial X_\mu}) \; , \qquad P_\mu = i\frac{\partial}{\partial X_\mu} \; ,$$

$$D = iX^\nu \frac{\partial}{\partial X_\nu} \; . \qquad (3\cdot2)$$

Now let Δ be a representation of G on a Hilbert space H, and λ a complete set of labels for a basis of H. The transformations $e^{i\alpha D}$ acting on f_λ will result, in general, in another function $f_{\lambda'}$, $\lambda' \neq \lambda$, or a combination of these functions. However, when α is restricted to the lattice $\alpha = 2\pi i n$ the labels λ will fail to distinguish between f_λ and $e^{-2\pi nD}f_\lambda$, since $e^{-2\pi nD} \simeq 1$ on H for the realization given by $(3\cdot2)$.

The easiest way to remove this degeneracy is to introduce a new fifth coordinate X_5 and change the realization of D into

$$D = i(X_\nu \frac{\partial}{\partial X_\nu} + \frac{\partial}{\partial X_5}) \qquad (3\cdot3)$$

(this changes the group structure but not the (CR) of the Lie algebra) and consider the functions

$$F_{\lambda,n} = \delta(X_5 - 2\pi in)f_\lambda \; . \qquad (3\cdot4)$$

Applying the operator $e^{-2\pi mD}$ to $F_{\lambda,n}$ we find

$$e^{-2\pi mD} \; F_{\lambda,n} \; = \; \delta\left[\lambda - 2\pi i(n+m)\right] \; f_{\lambda} \; = \; F_{\lambda,m+n} \; . \qquad (3 \cdot 5)$$

Thus a new quantum number is needed to specify completely the particles' wave functions. We note, however, that there exists no operator within our group structure such that $QF_{\lambda,n} = n \cdot F_{\lambda,n}$ and we must introduce such an operator externally. If we now define

$$J_{+} \; = \; e^{-2\pi D} \; , \qquad J_{-} \; = \; e^{2\pi D} \; , \qquad (3 \cdot 6)$$

we find that the three operators J_{+}, J_{-}, Q satisfy the following (CR)

$$\left[J_{+},Q\right] \; = \; - \; J_{+} \; , \qquad \left[J_{-},Q\right] \; = \; - \; J_{-} \; ,$$

$$\left[J_{+},J_{-}\right] \; = \; 0 \; . \qquad (3 \cdot 7)$$

From the physical point of view we know one operator which commutes with all the transformations of the Poincaré group, whose eigenvalues are multiples of a certain unit and which manifests a structure similar to that given by equation (3·7) (see next paragraph). This is the electric charge operator. We are led therefore to suggest the identification of Q with the electric charge.

Obviously the main objection to this identification is that the operators J_{+}, J_{-}, Q form an E(2) algebra rather than SU(2). To obtain this desired result we point out therefore that in both relativistic and non-relativistic physics the introduction of a non-zero charge changes some (CR) of the invariance group related to the translation operators. Thus in non-relativistic quantum mechanics with the Galilean invariance group we have

$$\left[P_{\mu},K_{\nu}\right] \; = \; m \, \delta_{\mu\nu} \; , \qquad (3 \cdot 8)$$

where m is the mass (gravitational charge) and P_{μ}, K_{ν} are the space and velocity translations. On the other hand it is well-known in relativistic physics that the introduction of an electromagnetic field changes the (CR) of the translation operators so that

$$[P_\mu, P_\nu] = iF_{\mu\nu}, \tag{3.9}$$

where $F_{\mu\nu}$ is the electromagnetic tensor. In view of these two examples it seems natural to assume that the electric charge translation operators J_+, J_- will in fact not commute but will satisfy

$$[J_+, J_-] = \alpha Q \tag{3.10}$$

which yields SU(2) as one of the possibilities.

IV. Conclusions

The main objection to our treatment might arise from the fact that we use the complex Poincaré group rather than the real one. We noticed already that this use is not new [6]. We remark moreover that the only place where complex transformations were used was in the quantization of the D operator. Thus once we perform this quantization and consider a particle with a definite charge we can restrict ourselves to real transformations only. Complex transformations are needed only to pass from a particle with one charge to another. An evidence that this is the real situation can be extracted from the fact that while the space-time symmetry groups (such as the Galilean and Poincaré groups) are real transformation groups, the inner symmetry groups, (such as SU(3)) which connect particles with different charges, are complex transformation groups.

As a further extension of the method presented in this paper we note that if we did not start by quantizing D then we should have the following (CR) between the mass spin and the dilatation operators:

$$[D, P^2] = P^2, \qquad [D, W^2] = W^2, \tag{4.1}$$

$$[P^2, W^2] = 0. \tag{4.2}$$

If we now require that a multiplet of particles form a projective representation [7] of this algebra (we must make this assumption since quantum mechanical states are phase invariant) then the (CR) between P^2, W^2 will change into

$$\left[P^2, W^2\right] = 1 . \qquad (4\cdot3)$$

We can now quantize $(4\cdot1)$, $(4\cdot2)$ simultaneously according to equations $(2\cdot6)$ and $(2\cdot2)$ respectively and find that the mass, spin and electric charge of the particles in the multiplet must be quantized [8]. Thus the mathematical method presented in section II affords a natural explanation for the quantization of the particles' charges in nature.

To conclude our discussion we note that classically physical observables are assigned to infinitesimal generators of the transformation group. Our treatment diverges basically from this approach since we tried to connect the electric charge to finite transformations on the global group manifold [9]. Such a basic difference might seem unjustified unless we note that the no-go theorems [10] exclude any physically reasonable unification of the space-time and inner symmetries within the classical scheme. In view of this situation our method might prove itself as a rewardable new approach to this problem and we hope to elaborate on it further in the future.

ACKNOWLEDGEMENTS

I am deeply indebted to Professor A. Grossman who called my attention to the papers of P. Cartier and J. Zak. The hospitality extended to me at the University of Toronto is gratefully acknowledged.

NOTES AND REFERENCES

1. E. C. Zeeman - J. Math. Phys. 5, p. 490 (1964).

2. J. Fulton, R. Rohrlich and L. Witten - Rev. Mod. Phys. 34 , p. 442 (1962).

3. P. Cartier - Symposia in Pure Math. Vol. 9, p. 361 (American Math. Soc.).

4. J. Zak - Phys. Rev. 168 , p. 686 (1968).

5. In other words this means that we can find a different Lie group which has the same Lie algebra.

6. A. O. Barut - in Lec. in Theo. Phys. 7A, p. 121 (1964) ed.

    W. E. Britten & A. O. Barut.

    - J. Math. Phys. 5, p. 1652 (1964).

7. V. Bargmann - Ann. of Math. 59, p. 1 (1954).

8. Here we emphasize that the restriction $\alpha\beta=2\pi n$ still allows a continuum of α,β solutions for every n. To overcome this we assume that for each n only one pair of (α,β) is chosen.

9. We remark also that our operators are nonunitary. Similar difficulties were encountred in other theories of the electric charge (R. J. Adler - J. Math. Phys. 11, p. 1185).

10. L. O'Raifeartaigh - Phys. Rev. 139 , p. 1052 (1965)

    S. Coleman - Phys. Rev. 138 , p. 1262 (1965) .

ON AN EINSTEIN-MAXWELL FIELD WITH A NULL SOURCE [*]

Lane P. Hughston [†]

The University of Texas, Austin, Texas ,

and

Massachusetts Institute of Technology, Cambridge, Massachusetts .

1. Introduction

A curious feature of Maxwell's theory is the admittance of fields for which the source is a <u>null current</u>, perhaps representing charge traveling with the velocity of light [1]. Let us say that an electromagnetic field F is that of <u>charged radiation</u> if F is null and the associated complex current J lies in the propagation direction:

$$J_{\wedge} (F + i\,^{*}F) = 0. \tag{1}$$

Assuming that the current has a nonvanishing vorticity,

$$J_{\wedge}\,^{*}dJ \neq 0, \tag{2}$$

it follows from Maxwell's equations [2]

$$J = -\delta(F + i\,^{*}F) \tag{3}$$

that there exists a null vector k tangent to an affinely parametrized diverging congruence of shearfree geodesic curves; k lying in the direction of J [3].

---

[*] Supported in part by N.S.F. Grants GP-8868, GP-20033 and GU-1598, Air Force Office of Scientific Research Grant AF-AFOSR-903-67 and NASA Grant NGL 44-004-001.

[†] Permanent address: 5314 Palomar Lane, Dallas, Texas 75229.

We would like to show that if F represents the class of all such fields in Minkowski space, then the restricted class thereof (for some curved spacetime metric preserving the nullity of J) satisfying Einstein's equations

$$\tfrac{1}{2} R = F \cdot F + {}^{*}F \cdot {}^{*}F ,$$ (4)

necessarily possesses a magnetic current as well as an electric current:

$$J \neq \overline{J}.$$ (5)

We thereby establish a plausible link between the absence of null currents with vorticity and the absence of magnetic currents in nature.

2. The Flat Background

As has been shown by Robinson and Robinson [4], coordinates $\{\zeta, \overline{\zeta}, \sigma, \rho\}$ may be chosen so that the general flat space-time line element built about the congruence assumes the form

$$ds_0^2 = 2(\rho^2 + \Omega^2) d\varsigma \, d\overline{\varsigma} + 2k(d\rho + Z d\varsigma + \overline{Z} d\overline{\varsigma} + Sk),$$ (6a)

$$k = b d\varsigma + \overline{b} d\overline{\varsigma} + d\sigma ,$$ (6b)

$$Z = \rho \Lambda - i(\Omega_1 + \Lambda \Omega) , \quad S = -\tfrac{1}{2}\left(\overline{\Lambda}_1 + \Lambda_2\right) ,$$ (6c)

$$b = \Lambda \left[\overline{\chi}(\overline{\varsigma}) - \sigma\right] , \quad \Omega = \tfrac{1}{2} i \left(\overline{b}_1 - b_2\right) ,$$ (6d)

$$\Lambda = \left[\varsigma + \overline{\lambda}(\overline{\varsigma})\right]^{-1} ,$$ (6e)

where for a function $f(\varsigma, \overline{\varsigma}, \sigma)$ we use the notation

$$df = f_1 d\varsigma + f_2 d\overline{\varsigma} + f_3 k ,$$ (7)

and where λ and χ are disposable functions, analytic in ς.

With respect to the flat background, the basic equations (1) and (3) reduce to

$$F = (v d\mathfrak{Z} + \bar{v} d\bar{\mathfrak{Z}})_{\wedge} k \quad , \quad v = v(\mathfrak{Z}, \bar{\mathfrak{Z}}, \sigma), \tag{8}$$

$$J = (v_{\mathfrak{2}} + \bar{\Lambda} v)(\rho^2 + \Omega^2)^{-1} k, \tag{9}$$

as may be seen by comparing Robinson, Schild and Strauss [5].

3. Gravitational Case

Let us ask that F should satisfy the basic equations with respect to a curved spacetime metric ds^2 as well as the background. As Robinson, Robinson, and Zund [6] have in effect commented, it follows as a necessary and sufficient condition that this metric should reduce to the Kerr-Schild form

$$ds^2 = ds_0^2 + 2\rho m(\rho^2 + \Omega^2)^{-1} kk, \quad m = m(\mathfrak{Z}, \bar{\mathfrak{Z}}, \sigma), \tag{10}$$

noting that k must be shearfree and geodesic with respect to ds^2 as well as ds_0^2.

Substituting expression (8) for the electromagnetic field into the gravitational equations (4), we are able to write

$$R = v\bar{v}(\rho^2 + \Omega^2)^{-1} kk \quad . \tag{11}$$

Algebraically specialspace-times subject only to equations (10) and (11), but allowing ds_0^2 to be any vacuum metric built about a basic vector field k, have been previously investigated [7]. These results indicate that the squared amplitude of the electromagnetic radiation is associated with the flux of m:

$$\frac{\partial m}{\partial \sigma} = \frac{1}{2} v\bar{v} \quad . \tag{12}$$

The remaining gravitational equations determine m explicity to be

$$m = -\frac{1}{6}\sigma^{-3}, \tag{13}$$

where we have fixed the origin of σ by writing

$$\Omega = \frac{1}{2} i (\Lambda_2 - \bar{\Lambda}_1) \sigma . \tag{14}$$

Thus, the general solution is given by

$$v = e^{i\theta} \sigma^{-2} \tag{15}$$

for an arbitrary real function $\theta(\zeta, \bar{\zeta}, \sigma)$ which determines the phase of the field. The field source is both an electric current and a magnetic current, the current being entirely electric only when $J = \bar{J}$.

In the electric case, we would then have to require that the phase be constant and that

$$e^{i\theta} \bar{\Lambda} = e^{-i\theta} \Lambda , \tag{16}$$

whence we would obtain

$$\lambda = e^{-2i\theta} \zeta \tag{17}$$

upon use of the definition (6e). However, a calculation establishes that this in turn implies

$$\Lambda_2 = \bar{\Lambda}_1 , \tag{18}$$

which states equivalently that the twist vanishes ($\Omega = 0$).

Thus, there exists no solution of the Einstein-Maxwell equations for which the electromagnetic field is twisting radiation with an electric null current as a source, and the background gravitational field is flat.

The author wishes to express his gratitude to J. Ehlers, I. Robinson, and M. Walker for helpful conversations.

REFERENCES

1. W. B. Bonnor, Int. J. Theor. Phys. 2, 373 (1969); P. Dolan, Nature 227, 835 (1970).

2. I. Robinson, J. Math. Phys. 2, 290 (1961); M. Chevalier, Comp. Rend. A263, 526 (1966).

3. R. K. Sachs, Proc. Roy. Soc., A264, 309 (1961).

4. I. Robinson and J. R. Robinson, Int. J. Theor. Phys. $\underline{2}$, 231 (1969);

    M. Cahen and J. Leroy, J. Math. and Mech., $\underline{16}$, 501 (1966).

5. I. Robinson, A. Schild, and H. Strauss, Int. J. Theor. Phys. $\underline{2}$, 243 (1969).

6. I. Robinson, J. R. Robinson, and J. D. Zund, J. Math. and Mech., $\underline{18}$, 881

    (1969).

7. L. P. Hughston, Int. J. Theor. Phys., $\underline{4}$, 25 (1971).

THE LUMINOSITY OF A COLLAPSING STAR

Pt. I. H. Dwivedi and R. Kantowski

Department of Physics, University of Oklahoma

I. Introduction

Previous calculations of a collapsing star's luminosity have avoided the
difficulties of geometrical optics by resorting to Liouville's theorem of
relativistic kinetic theory   [1],[2]. In this paper we integrate the optical
scalar equations and apply their solutions to the collapsing star problem. We
obtain the corrected luminosity of reference[2].

The procedure is to first find the intensity of a point source with intrinsic
luminosity δL in a Schwarzschild field as seen by a distant observer, see Fig. 1.
This is done by substituting into [3]

$$\delta \mathcal{X} = \delta L \frac{\delta \Omega_s}{4\pi A_o \ (1+z)^2} \ , \tag{1}$$

where A_o is the area of a small beam of light at the observer, $\delta \Omega_s$ the solid angle
of the beam at the point source, and z is the red shift from source to observer.
Once we have this intensity we calculate the luminosity of an entire star by
assuming that its surface is covered with incoherent point sources and simply
integrating their intensities.

II. A Point Source

In equation (1) the area A is related to the expansion of the light beam
by   [4]

127

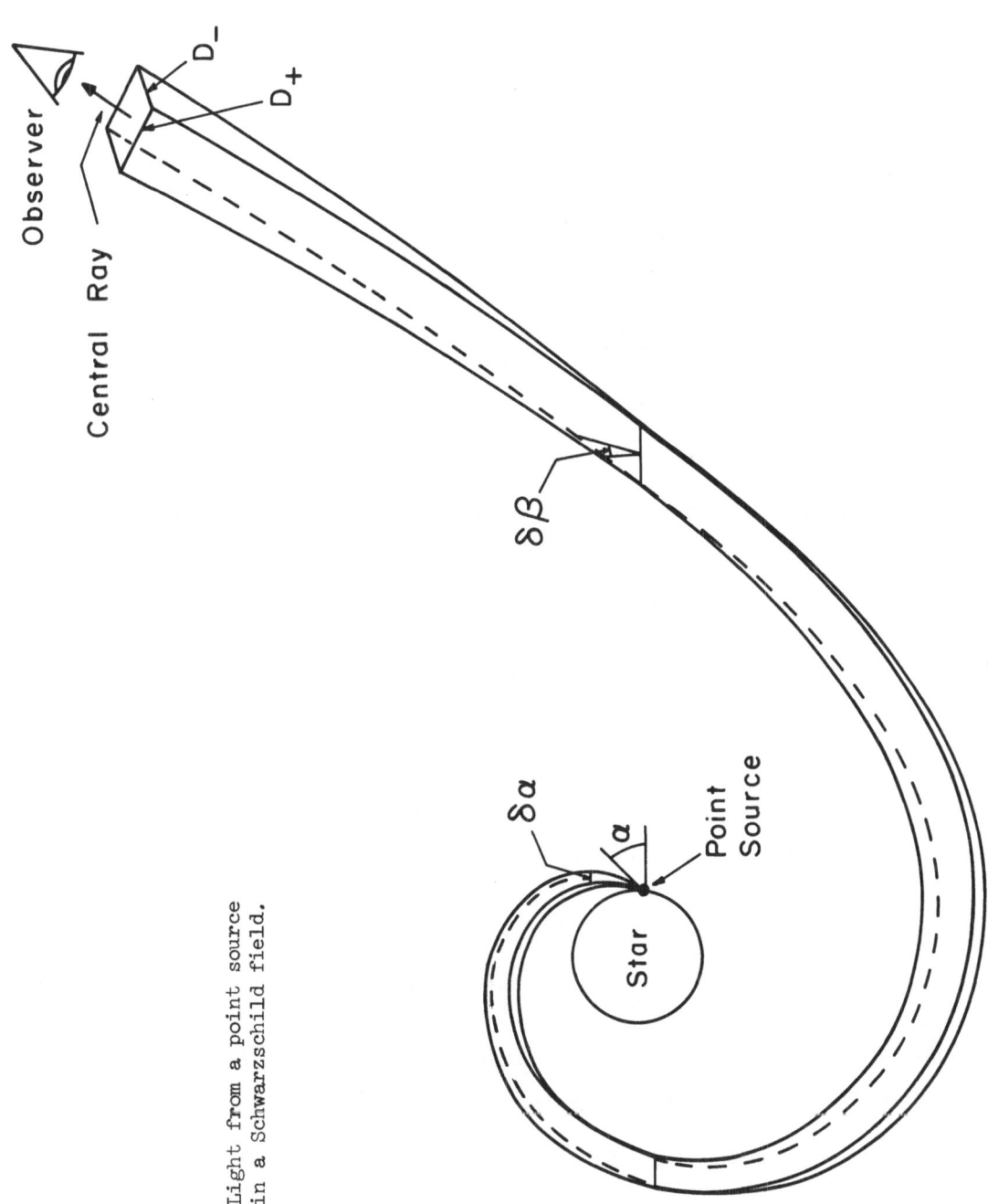

Figure 1.   Light from a point source
in a Schwarzschild field.

$$\dot{A}/A = 2\theta \tag{2}$$

and θ is calculated by integrating the optical scalar equations along the central light ray of beam [5,6]:

$$\dot{\theta} + \theta^2 + \sigma\bar{\sigma} = 0, \tag{3}$$

$$\dot{\sigma} + 2\sigma\theta = 3\,\frac{mk^2 \ell^2}{r^5}, \tag{4}$$

as usual, m is the Schwarzschild mass, k and ℓ are integration constants along the central ray (see equations (13) - (16)), and (') is differentiation with respect to the central ray's affine parameter. The phase in the driving term of equation (4) has been adjusted so that the shear, σ, will remain real if it starts that way. The above equations decouple if we rewrite the principal curvatures ($K_{\pm} = \theta \pm \sigma$) [7] in terms of the major and minor dimensions of the wave front, $D_{\pm}$ (see figure 1), i.e.,

$$K_{\pm} = \frac{\dot{D}_{\pm}}{D_{\pm}}. \tag{5}$$

Rewriting equations (3) and (4) we have

$$\ddot{D}_{\pm} \mp 3\frac{mk^2 \ell^2}{r^5}\,D_{\pm} = 0, \tag{6}$$

whose solution when evaluated at the observer gives us $A_o = D_+ D_-$, and hence the luminosity of a point source. We find D_- by using the following theorem: If a point source has a one parameter isotropy group generated by a hypersurface orthogonal killing vector ξ_a then $\xi_a/|\xi|$ is a parallelly transported principal shear direction whose principal curvature is given by $K_- = |\dot{\xi}|/|\xi|$.

The proof of this theorem is straightforward; you take the central null ray k_a, Lie transport it using,

$$k_{a;b}\,\xi^b = \xi_{a;b}\,k^b, \tag{7}$$

and then use the hypersurface orthogonality,

$$\zeta_{a;b} = \xi_{[a} \, (\ln \xi^2)_{,\,b]} \, , \tag{8}$$

to get

$$k_{a;b} \, \xi^b = \xi_a \, |\dot{\xi}| / |\xi| \, . \tag{9}$$

This says that ξ^a is an eigen-direction of $k_{a;b}$ with eigenvalue $|\dot{\xi}|/|\xi|$, i.e., a principal curvature direction with principal curvature $|\dot{\xi}|/|\xi|$. It is also straightforward to prove that ξ^a is space-like and its direction parallelly transported.

For Schwarzschild the isotropy group is the set of rotations which leave the source fixed. If we put the source at $r_s = R > 2m$, $\phi_s = 0$, $\theta_s = 2\pi$ we have

$$\zeta = \sin\phi \, \frac{\partial}{\partial\theta} + \cot\theta \, \cos\phi \, \frac{\partial}{\partial\phi} \, , \tag{10}$$

and

$$|\zeta| = r \left[\sin^2\phi + \cos^2\phi \, \cos^2\theta \right]^{\frac{1}{2}} \, . \tag{11}$$

By orienting the coordinates so that the central ray is in the $\theta = \pi/2$ plane we have $|\xi| = r \sin\phi$ and hence

$$D_- = \text{con.} \, (r \sin\phi) = \delta\beta \, r \sin\phi \, , \tag{12}$$

where β is the isotropy group parameter specified by $\sin\beta = (1 + \tan^2\theta \sin^2\phi)^{-\frac{1}{2}}$ (see fig. 1).

To calculate D_+ we must go directly to the Schwarzschild case. Let $x^a(\lambda, \beta, \ell)$ be the null hypersurface (optical pulse), where λ is the affine parameter, β the isotropy parameter, and ℓ the impact parameter at infinity [2] (see fig. 2).

The tangents to the null geodesics making up the hypersurface are

$$k^t = k/\left(1 - \frac{2m}{r}\right) , \tag{13}$$

$$k^r = k\left\{1 - \frac{\ell^2}{r^2} \left(1 - \frac{2m}{r}\right)\right\}^{\frac{1}{2}} , \tag{14}$$

$$k^\theta = \frac{k\ell}{r^2} \cos\phi \, \sin\beta \, , \tag{15}$$

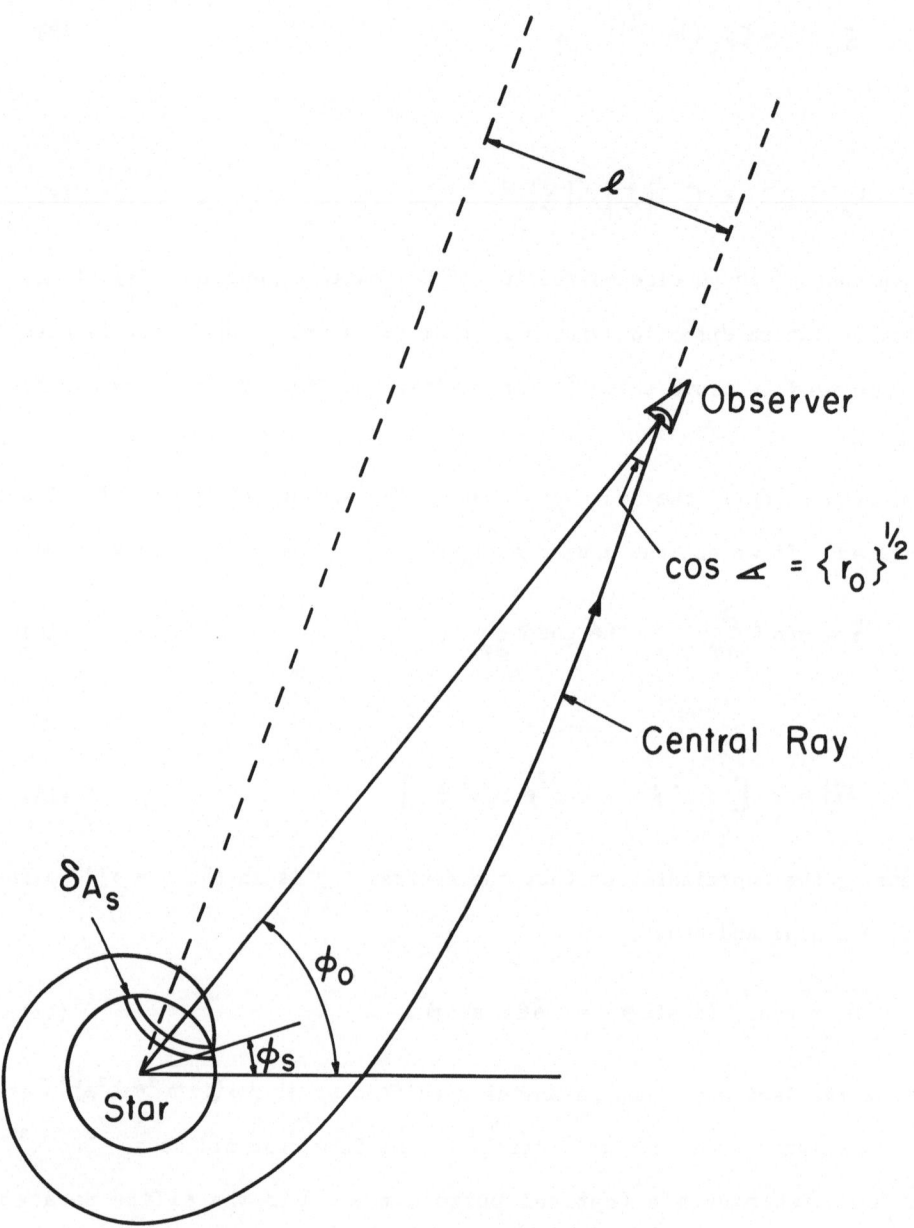

Figure 2. Calculating δA_s on the star's surface.

$$k^{\phi} = \frac{k\ell}{r^2} \cos\beta / \sin^2\theta \quad . \tag{16}$$

We calculate D_+ by calculating the magnitude of $\dfrac{\partial x^{\alpha}(\lambda, \rho, \ell)}{\partial \ell}$,

$$\frac{\partial x^a}{\partial \ell} = \frac{\partial}{\partial \ell} \int k^a d\lambda = \frac{\partial}{\partial \ell} \int_R^r \frac{k^a}{k^r} dr \quad . \tag{17}$$

This is straightforward and when evaluated along the central curve gives

$$D_+ = \delta\ell\, r\{r\}^{1/2} \int_R^r \frac{dr}{r^2 \{r\}^{3/2}} \quad , \tag{18}$$

where

$$\{r\} \equiv \left\{ 1 - \frac{\ell^2}{r^2} \left(1 - \frac{2m}{r} \right) \right\} \quad ,$$

and will be used throughout.

It is easy to verify that both $D_{\pm}$ satisfy equation (6). If we were interested in the luminosity of a point source rather than a star we would now put both $D_{\pm}$ into equation (1) along with the solid angle [8]

$$\delta\Omega_s = \sin\alpha \,\, \delta\alpha \,\, \delta\beta \tag{19}$$

and have the intensity of a point source once we evaluated the redshift. The result would be given in terms of the observers coordinates r_0, ϕ_0 and R, ℓ, α ; not all of which are independent, i.e.,

$$\sin\alpha = \frac{\ell}{R} \left(1 - \frac{2m}{R} \right)^{1/2} \tag{20}$$

and

$$\phi_0 - \phi_s = \int_R^{r_0} \frac{dr}{r^2 \{r\}^{1/2}} \quad . \tag{21}$$

III. The Star

If L is the intrinsic luminosity of the star we can write the equivalent point source luminosity of a small piece of its surface area, δA_s, as

$$\delta L = \frac{2L}{4\pi R^2} \,\, \delta A_s \quad , \tag{22}$$

where L and R are evaluated at the retarded time t_s that δA_s emits the observed light. In these calculations, in contrast to those of the previous section, we are keeping the observer's end of the null ray fixed ($t = t_0$, $r = r_0$, $\phi = \phi_0$, $\theta = \pi/2$) and are moving around to different points on the star's surface, i.e., different R, t_s, ϕ_s, and θ_s. These parameters are, of course, all fixed by giving ℓ, β, and the equation of the star's collapsing surface R(t). We will assume the star collapses roughly as free fall, i.e., has a reasonable Schwarzschild velocity v at R = 3m and continues on through R = 2m,

$$\dot{R} \equiv \frac{dR}{dt} = \left(1 - \frac{2m}{R}\right) v, \text{ where } v \sim 1. \tag{23}$$

Using this to integrate equations (13) and (14) we get $t_s(\ell)$ and R (ℓ),

$$t_0 - t_{3m} = t_s - t_{3m} + \int_{R\,(t_s)}^{r} \frac{dr}{\left(1 - \frac{2m}{r}\right)\{r\}^{1/2}} \qquad , \tag{24}$$

where t_{3m} is the time the star crosses R = 3m, and where

$$t_s - t_{3m} = \int_{3m}^{R(t_s)} \frac{dR}{\dot{R}} = 2m \ln \left(\frac{R}{m} - 2\right) + 0 \text{ (m)}. \tag{25}$$

We will only need those values of ϕ_s in the $\theta = \pi/2$ plane and they are given by equation (21).

From the rotational symmetry about the axis from the observer to the star's center we can use concentric rings for δA_s (see figure 2),

$$\delta A_s = 2\pi R^2 \sin (\phi_s - \phi_0) \, \delta \phi_s \qquad , \tag{26}$$

with equivalent point source brightness

$$\delta L = L \sin (\phi_s - \phi_0) \, \delta \phi_s. \tag{27}$$

Putting this and $D_{\pm}$ (where $\phi \rightarrow \phi_0 - \phi_s$) into the luminosity equation (1), writing $\alpha = \alpha(\ell, R)$ by equation (20), correcting for the cosine of the angle between the observer's screen and the incoming photons ($\{r_0\}^{1/2}$), and integrating we have the luminosity of a collapsing star

$$\mathcal{L} = \int \frac{\frac{L \sin\alpha}{2} \left.\frac{\partial\alpha}{\partial\ell}\right|_R \; \delta\phi_s}{4\pi r_o \int_R^{r_o} \frac{dr}{r^2\{r\}^{3/2}}} \cdot (1+z)^2 \qquad . \tag{28}$$

The terms in this integral are easy to evaluate, the α term from equation (20) and

the redshift is

$$1 + z = \frac{1 + v\{R\}^{1/2}}{\left[1- v^2\right]^{1/2}} \; \frac{\left(1- \frac{2m}{r_o}\right)^{1/2}}{\left(1 - \frac{2M}{R}\right)^{1/2}} \qquad , \tag{29}$$

where the observer is assumed at rest in the Schwarzschild coordinates. The

integration over ϕ_s would be several multiples of 2π (in the late stages of

collapse) due to some photons circling the star several times near $r = 3m$ [2].

The integration is, however, best performed by using ℓ, the impact parameter at

infinity. The appropriate substitution gives

$$\delta\phi_s = \left.\frac{\partial\phi}{\partial\ell}\right|_{r_o,t_o} \; d\ell \quad , \tag{30}$$

where

$$\left.\frac{\partial\phi}{\partial\ell}\right|_{r_o,t_o} = \int_R^{r_o} \frac{dr}{r^2\{r\}^{3/2}} \left[\frac{1 + v\{R\}^{1/2}}{\{R\}^{1/2} + v}\right]^{1/2} \{R\}^{1/2} \quad . \tag{31}$$

The luminosity is then given by

$$\mathcal{L} = \frac{1}{4\pi r_o^2} \int_0^{\ell_{max}} \frac{L\left(\frac{\ell}{R}\right)\left(1- \frac{2m}{R}\right)^2 [1-v^2] \; d\ell}{\left(1- \frac{2m}{r_o}\right)\left[\{R\}^{1/2}+ v\right]\left[1+v\{R\}^{1/2}\right]} \quad . \tag{32}$$

To evaluate the integral we must find ℓ_{max}, $L(\ell)$, and $R(\ell)$. The ℓ dependence of

R comes from equations (23)-(25), and in the late stages of collapse we have

given its typical behavior in Figure 3. The ℓ dependence of the stars luminosity

L occurs because of its time dependence and is whatever we wish; however, as long

as it doesn't change significantly on time scales $\sim m$, it can be taken as a

constant. If we put $\alpha = \pi/2$ in equation (20) we have ℓ_{max}. In the early stages

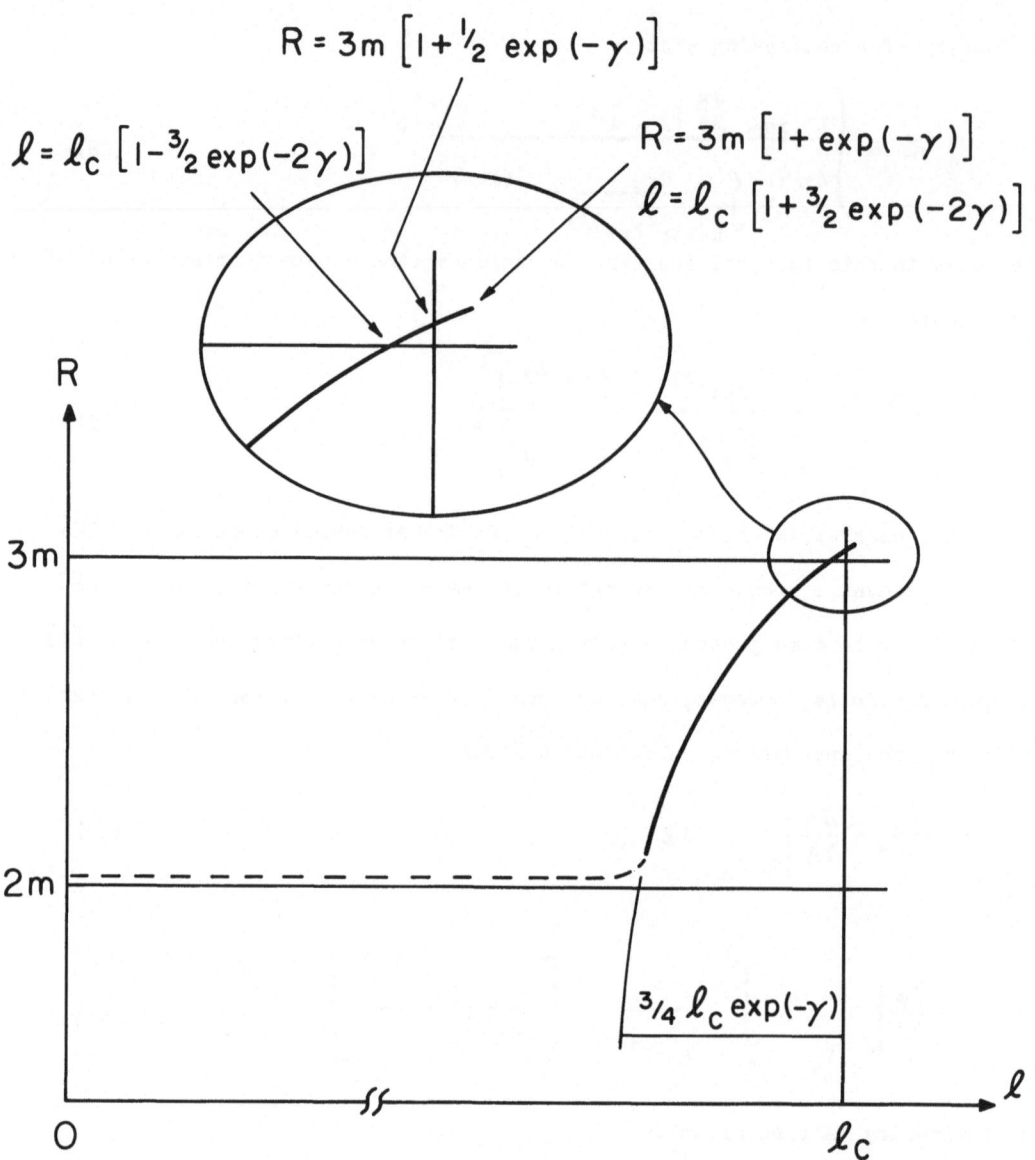

Figure 3. R(ℓ) curve for typical collapsing star given analytically for R$\not=$2m by

$$R \sim \frac{3m}{2} \frac{1 + \left[1 - \frac{2}{3}\left(1 - \frac{\ell^2}{\ell_c^2}\right) \exp(+\gamma)\right]^{\frac{1}{2}}}{1 - \frac{1}{2}\exp(-\gamma)} ,$$

where $\gamma \equiv \dfrac{t_o - r_o - t_{3m}}{3\sqrt{3}\,m}$ and $\ell_c \equiv 3\sqrt{3}\,m$.

of collapse where m/R<<1, we see the luminosity reduces to $L/4\pi r_o^2$ as expected (assume v~0). In the late stages of collapse, however, (i.e., after the star has gone through its Schwarzschild radius) we find an exponentially decaying luminosity. To see this first observe that the $(1-\frac{2m}{R})^2$ term in equation (32) makes the luminosity due to light emitted from R~2m negligible. Next observe from the R(ℓ) curve in Figure 3 that R changes from R~2m to ~3m rather abruptly in a narrow time dependent region $\Delta\ell$ where ℓ~$\ell_c \equiv 3\sqrt{3}m$. This, when put into equation (32) shows that the major part of the luminosity is due to radiation emitted when the star went through R~3m and appears to be coming from a ring on the edge of the star. The width of the ring $\Delta\ell$, decays exponentially and gives an exponentially decaying total luminosity for the star. The approximate analytical expression for $\mathcal{L}$, correct to an order of magnitude, is

$$\mathcal{L} = \frac{L}{4\pi r_o^2} \; \frac{[1-v_{3m}^2]}{9 v_{3m}} \; \exp\left\{-\frac{(t_o - r_o - t_{3m})}{3\sqrt{3}\,m}\right\} \; , \tag{33}$$

where v_{3m} is the star's velocity as seen by a rest observer in Schwarzschild at R = 3m. In this expression we have put $2m/r_o = 0$ and have neglected terms of order m in the exponential. The somewhat mysterious v_{3m} in the denominator is due to the fact that the number of photons emitted into the R ~3m region when the star went through is proportional to L/v_{3m}. The photons present in this region then exhibit the characteristic exponential escape and are redshifted from source to observer giving equation (33).

REFERENCES

1. M. A. Podurets, Astr. Zh. **41**, 1090 (1964), (English trans. Soviet Ast.-AJ. **8**, 868 (1965).

2. W. L. Ames and K. S. Thorne, Ap. J. **151**, 659 (1968).

3. J. Kristian and R. K. Sachs, Ap. J. **143**, 379 (1966).

4. R. K. Sachs, Proc. Roy. Soc. London, A, **264**, 309 (1961).

5. R. Kantowski, Ap. J. **155**, 89 (1969).

6. We are using Schwarzschild coordinates and units $c = G = 1$ throughout.

7. R. Kantowski, J. Math. Phys. 9, 336 (1968).

8. The angle α is defined in Figure 1 (see equation (20)).

A CLASS OF INEXTENDIBLE WEYL SOLUTIONS[*]

Burton H. Voorhees

Center for Relativity Theory
University of Texas at Austin
Austin, Texas

INTRODUCTION

The Weyl solutions are those solutions of the vacuum Einstein equations which are static and axially symmetric [1]. That is, they possess two Killing vectors, one time-like with orbits $\mathbb{R}$ and the other space-like with orbits S^1.

The purpose of this paper is to study the possibility of obtaining extensions for a subclass of the Weyl solutions which contains as special members the Schwarzschild and Curzon solutions. It will be shown that only the Schwarzschild solution is extendible.

EXTENSIONS

Extension of a spacetime manifold should be viewed as follows: For a spacetime M choose coordinates such that the TL Killing vector is

$$\xi = \frac{\partial}{\partial t} \qquad . \qquad (1)$$

Then by definition, the norm of ξ is

$$\phi = (-g_{ab}\xi^a\xi^b)^{1/2} = (-g_{44})^{1/2} \qquad . \qquad (2)$$

In general, there may be a horizon. That is, a hypersurface Σ which is a boundary of M characterized by $\phi|_{\Sigma} = 0$. Suppose that ℓ^a is the tangent vector

---

[*] Based in part on a Dissertation submitted to the University of Texas at Austin, 1971.

to some TL geodesic γ intersecting Σ. Then $\xi_a \ell^a = -E$ is constant along γ, and $\dot{t} = E/\phi^2$ is a first integral of the geodesic equations. On Σ, however, this becomes unbounded. For some finite value of proper time, the coordinate t becomes infinite; hence, the geodesic is incomplete at Σ.

If M is extendible, it is possible to attach a new four-dimensional region M' onto M along Σ such that if geodesics in M which intersect Σ are continued into M' they are either complete or reach some point of M' beyond which they are inextendible (i.e., they encounter a singularity). The simplest example is the Kruskal extension of the Schwarzschild manifold[2](Figure 1). The horizon is the line r = 2m and there is a singularity at r = 0. In the original Schwarzschild coordinates (r, θ, φ, t), the metric is singular at the horizon and TL geodesics are incomplete there. In the extended solution, these geodesics may be continued until they reach r = 0 beyond which extension is impossible.

The usual method of proving that a manifold is inextendible involves demonstrating that almost all geodesics are either complete or inextendible.

THE CLASS OF SOLUTIONS

The metrics to be considered are

$$ds^2 = \kappa^2 \left(\frac{\cosh \eta + 1}{\cosh \eta - 1}\right)^\delta \left\{ (\cosh^2 \eta - \sin^2 \xi) \frac{\sinh^2 \eta}{\cosh^2 \eta - \sin^2 \xi} \right.$$

$$\left. \times (d\eta^2 + d\xi^2) + \sinh^2 \eta \cos^2 \xi \, d\varphi^2 \right\} - \left(\frac{\cosh \eta - 1}{\cosh \eta + 1}\right)^\delta dt^2 , \quad (3)$$

with $\delta = m/\kappa$, m being the Bondi mass, and (η, ξ) are spheroidal coordinates defined in terms of Weyl's canonical cylindrical coordinates by

$$\rho + iz = \kappa \sinh(\eta + i\xi), \quad \eta > 0, \quad -\frac{\pi}{2} \leq \xi \leq \frac{\pi}{2} . \quad (4)$$

In the limit $\delta \to \infty$ (m remaining finite), (3) becomes the Curzon metric, and if $\delta = 1$, the transformation $\cosh \eta \to r/m - 1$, $\sin \xi \to \cos \theta$ shows that (3)

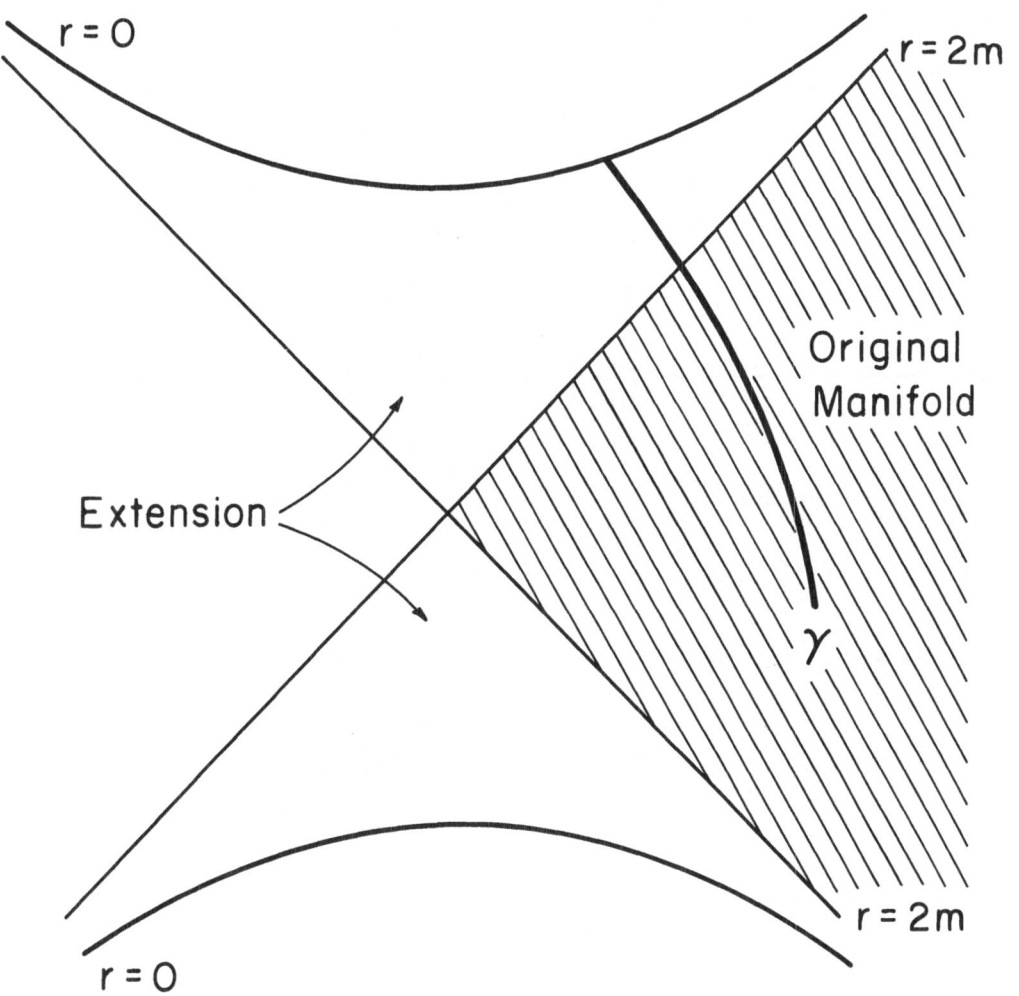

Figure 1

is the Schwarzschild solution.

In all cases, there is a horizon at $\mathcal{N} = 0$, and if $\delta \neq 1$, there is a singularity at $\mathcal{N} = 0$ ($\xi \neq \pm\frac{\pi}{2}$).

The discussion of singularities is best carried on in terms of the divergences of scalar invariants of the curvature tensor. One such invariant which is particularly simple for the Weyl solutions is [3]

$$\alpha = \frac{e^{-2(\nu-\gamma)}}{R_0^2(\cosh^2\mathcal{N} - \sin^2\xi)}\left\{\left(\frac{\partial\gamma}{\partial\mathcal{N}}\right)^2 + \left(\frac{\partial\gamma}{\partial\xi}\right)^2 - \frac{\partial\gamma}{\partial\mathcal{N}}\coth\mathcal{N} + \frac{\partial\gamma}{\partial\xi}\tan\xi\right\} \quad (5)$$

for a metric of the form

$$ds^2 = e^{2(\nu-\gamma)}(d\mathcal{N}^2 + d\xi^2) + \rho^2 e^{-2\gamma}d\varphi^2 - e^{2\gamma}dt^2 \quad . \quad (6)$$

In the case being considered,

$$\alpha = \frac{\delta}{\kappa^4}\left(\frac{\cosh\mathcal{N} - 1}{\cosh\mathcal{N} + 1}\right)^\delta \frac{(\cosh^2\mathcal{N} - \sin^2\xi)^{\delta^2-1}(\cosh\mathcal{N} - \delta)}{(\sinh^2\mathcal{N})^{\delta^2+1}} \quad . \quad (7)$$

If $\delta \neq 1$, then α diverges at $\mathcal{N} = 0$ with the exception that for $\delta \geq 2$, $\xi = \pm\frac{\pi}{2}$ α becomes zero at $\mathcal{N} = 0$. Similar results have been discussed by Gautreau and Anderson [4]. This is the so-called directional singularity, for if $\mathcal{N} = 0$ is approached along the symmetry axis no singularity is observed.

RESULTS

The main result of this paper may be stated as a theorem:

"The only extendible solution in the class of

metrics (3) is the Schwarzschild solution."

The proof of this proceeds through two stages:

Lemma I:

The metrics (3) are geodesically complete at spatial infinity.

Proof:

In a previous paper [5] I have shown that if the metrics (3) are transformed to Schwarzschild coordinates (r, θ, φ, t) and expanded in powers of r^{-1} then to first order they are identical to the Schwarzschild metric. Since the Schwarzschild metric is geodesically complete at infinity, the lemma follows.

Lemma II:

The only incomplete, extendible geodesics in a spacetime with metric (3) are those contained in the symmetry axis, which for $\delta \geq 2$ is an extendible 2-surface.

Proof:

That the symmetry axis is an extendible totally geodesic 2-surface if $\delta \geq 2$ has been demonstrated by Walker (private communication) and by Godfrey [6]. Also if $\delta < 2$, it is clear that no extension is possible.

Consider some TL geodesic contained in the symmetry 2-surface having tangent vector T^a. Along this geodesic, construct a Jacobi field ζ^a connecting this geodesic to a neighboring geodesic (Figure 2). For simplicity, assume $\zeta^\varphi = 0$ (this does not change the final results).

By construction, $T^a = (\dot{\eta}, 0, 0, \dot{t})$ and $\zeta_a T^a = 0$. Since T^a is geodesic, $\dot{\zeta}_a T^a = \ddot{\zeta}_a T^a = 0$ where the dot is defined by $\dot{\lambda}^a = T^b \nabla_b \lambda^a$. Using $\zeta \cdot T = -E$, we obtain from the geodesic equations

$$\dot{\eta} = e^{-\nu}(E^2 - e^{2\gamma})^{1/2} \quad , \quad (8)$$

$$\dot{t} = -e^{-2\gamma}E \quad ,$$

and from the equation of geodesic deviation

$$\ddot{\zeta}^\eta = e^{-2(\nu-\gamma)}\left[\left(\frac{\partial \nu}{\partial \eta} - \frac{\partial \gamma}{\partial \eta}\right) - E^2 e^{-2\gamma}\frac{\partial \nu}{\partial \eta}\right] \quad ,$$

$$\ddot{\zeta}^\zeta = e^{-2(\nu-\gamma)}\left[\left(\frac{\partial \nu}{\partial \zeta} - \frac{\partial \gamma}{\partial \zeta}\right) - E^2 e^{-2\gamma}\frac{\partial \nu}{\partial \zeta}\right] \quad . \quad (9)$$

For the metric (3), equations (8) and (9) may be integrated near $\eta = 0$ to obtain

142

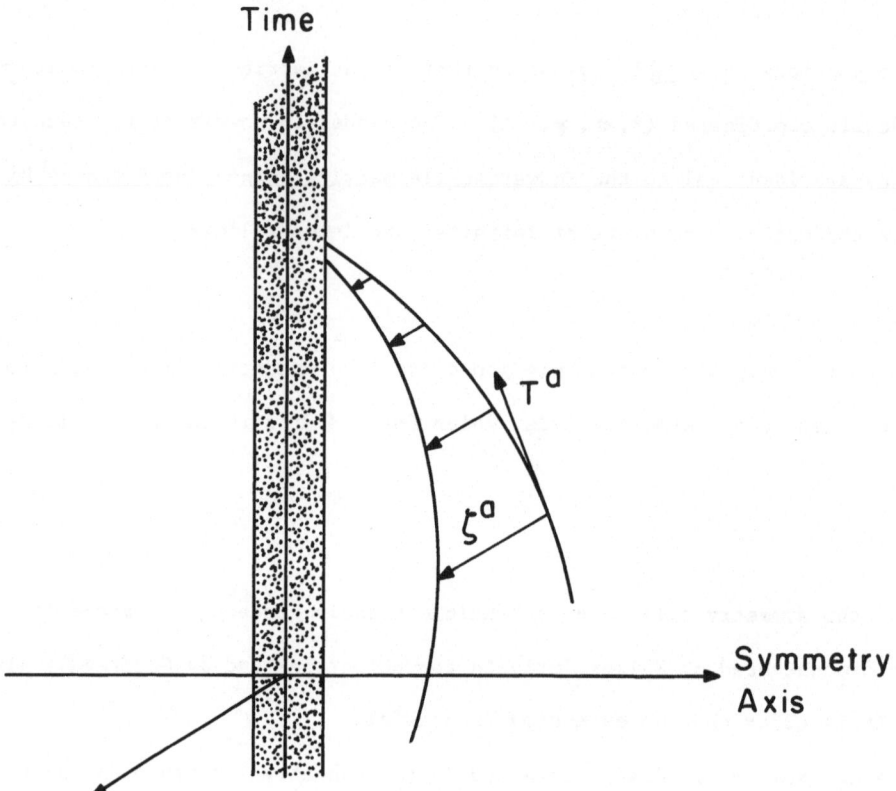

Figure 2

$$\eta \sim \left(\frac{2E}{\kappa}\right)s \quad ,$$

$$\ddot{\zeta}^\eta \sim -\frac{1}{2}\left(\frac{\kappa}{2E}\right)s^{-3/2} \quad ,$$

$$\dot{\zeta}^\eta \sim \left(\frac{\kappa}{2E}\right)s^{-1/2} + B_0 \quad ,$$

$$\zeta^\eta \sim 2\left(\frac{\kappa}{2E}\right)s^{1/2} + B_0 s + B_1 \quad ,$$

$$\dot{\zeta}^\xi = A_0 \quad ,$$

$$\zeta^\xi = A_0 s + A_1 \quad , \tag{10}$$

and $A_1 = 0$ for the case in question. Now the quantities $\zeta_a \zeta^a$ and $\dot{\zeta}_a \dot{\zeta}^a$ are computed using $\zeta_a T^a = \dot{\zeta}_a T^a = 0$ to eliminate terms in $\dot{\zeta}^t$:

$$\zeta_a \zeta^a \sim \frac{B_0^2 \kappa^2}{E^2} \eta^2 + \frac{A_0^2 \kappa^2}{4E^2} \eta^{6-2\delta} + 0(\eta^3) \quad ,$$

$$\dot{\zeta}_a \dot{\zeta}^a \sim \frac{\kappa^2}{E^2} + 2^\delta A_0^2 \kappa^2 \eta^{2(1-\delta)} + 0 \, (\eta) \quad . \tag{11}$$

If $\delta > 3$, both of these are singular at $\eta = 0$, and for $\delta \geq 2$, $\dot{\zeta}_a \dot{\zeta}^a$ always diverges at $\eta = 0$. The exception in either case is if $A_0 = 0$. However, $A_0 = 0$ means $\zeta^\xi = 0$, and hence, both geodesics are contained within the symmetry axis. This finishes the proof.

DISCUSSION

The results obtained point out the danger of asserting that every manifold with an extendible 2-surface is extendible as a manifold.

Also, it should be possible to generalize these results. A conjecture of Walker and the author is that the necessary and sufficient condition for a static asymptotically flat spacetime (possessing a horizon) to be extendible is that it be algebraically special.

I would like to thank M. Walker for several helpful suggestions and J. Ehlers for discussing many aspects of this problem.

REFERENCES

1. H. Weyl, Ann. Phys. $\underline{54}$, 47 (1917).

2. M. Kruskal, Phys. Rev. $\underline{119}$, 1743 (1960).

3. P. Jordan, J. Ehlers and W. Kundt, Abhandl. Akad. Wiss. Mainz, Math.-Nat.-Kl., Nr. 2, 1960.

4. R. Gautreau and J. L. Anderson, Phys. Letters $\underline{25A}$, No. 4, 291 (1967).

5. B. H. Voorhees, Phys. Rev. D $\underline{2}$, No. 10, 2119 (1970).

6. B. Godfrey, Unpublished Dissertation, Princeton University (1970).

SCALING IN FUNCTION SPACES

Jeffrey Winicour

Aerospace Research Laboratories

Wright-Patterson AFB, Ohio 45433

I. Scale Transformations

This paper concerns the subgroup of conformal transformations of a Riemannian metric,

$$g_{\mu\nu} \;\longrightarrow\; \lambda^2 \, g_{\mu\nu} \, .$$

for which the conformal factor λ is a constant. I shall refer to this simplest type of conformal transformation as a scale transformation. Because the scale factor λ is constant, scale transformations preserve not only the Weyl tensor,

$$C^{\mu}{}_{\nu\rho\sigma} \;\longrightarrow\; C^{\mu}{}_{\nu\rho\sigma} \, ,$$

but also the Riemann and Einstein Tensors,

$$R^{\mu}{}_{\nu\rho\sigma} \;\longrightarrow\; R^{\mu}{}_{\nu\rho\sigma} \, ,$$

$$G_{\mu\nu} \;\longrightarrow\; G_{\mu\nu} \, ,$$

so that they map vacuum solutions of Einstein's equations into vacuum solutions. Consequently, scale transformations provide a natural decomposition of the "space" of space-times into rays corresponding to scale related geometries. Given the physical and geometrical properties of some particular space-time, it is fairly straightforward to apply the appropriate scaling laws to uncover the physical and geometrical properties of all other space-times lying along the same ray.

In general, a ray contains a one-parameter set of inequivalent space-times. In special cases of scale symmetry, however, a ray consists of only one distinct geometry. A scale symmetry implies the existence of only one distinct geometry. A scale symmetry implies the existence of a one-parameter family of maps S_λ of points p of the manifold into points P_λ,

$$S_\lambda : p \to P_\lambda ,$$

such that the action on the metric of the induced map $S_{\lambda*}$ of the tangent spaces satisfies

$$g_p \to \left(S_{\lambda*} g \right)_{P_\lambda} = \lambda^2 g_{P_\lambda} .$$

For instance, the dilatations of flat space, represented in rectilinear coordinates by

$$x^\mu \to \lambda^{-1} x^\mu , \quad \eta_{\mu\nu} = diag \left(1, -1, -1, -1 \right) \to \lambda^2 \eta_{\mu\nu} ,$$

constitute a scale symmetry. This reflects the fact, which is obvious from the scaling behavior of the Riemann tensor, that flat space constitutes an entire ray of geometries.

It is easy to see that non-flat complete space-times with bounded curvature properties cannot possess a scale symmetry. Along a trajectory of such a symmetry, the scalar components of the curvature tensor have the monotonic dependence

$$\left[R_{\mu\nu\rho\sigma} A^\mu B^\nu C^\rho D^\sigma \right]_p = \lambda^{-2} \left[R_{\mu\nu\rho\sigma} A^\mu B^\nu C^\rho D^\sigma \right]_{P_\lambda} .$$

Consequently, given any local experiment which measures gravitational tidal forces (curvature), a point in space-time exists for which the results are larger (or smaller) than any specified value.

These considerations indicate that scale symmetries are not an expected feature of non-flat physically realistic localized systems. However, the concept of scale transformations which transform only the geometry and not the underlying manifold can be applied to any space-time. What makes this simple concept interesting is that it appears to have some non-trivial implications for the neighborhood of flat space in the space of space-times (see Sec. IV).

II. Energy Scaling

The first attempt to apply scale transformations to shed light on an important physical question was made by Einstein [1] in his 1941 paper entitled "Demonstration of the Non-Existence of Vacuum Gravitational Fields with Non-Vanishing Total Mass Free of Singularities". Einstein based his analysis on the observation that the Schwarzschild metrics with mass m_1,

$$ds_1^2 = \left(1 - \frac{2m_1}{r_1}\right)dt_1^2 - \left(1 - \frac{2m_1}{r_1}\right)^{-1}dr_1^2 - r_1^2\left(d\theta_1^2 + \sin^2\theta_1\, d\phi_1^2\right) ,$$

and mass m_2,

$$ds_2^2 = \left(1 - \frac{2m_2}{r_2}\right)dt_2^2 - \left(1 - \frac{2m_2}{r_2}\right)^{-1}dr_2^2 - r_2^2\left(d\theta_2^2 + \sin^2\theta_2\, d\phi_2^2\right),$$

are related by the scale transformation

$$ds_1^2 = \left(\frac{m_1}{m_2}\right)^2 ds_2^2 \quad .$$

This is easily seen by carrying out the coordinate transformation

$$t_1 = \frac{m_1}{m_2} t_2 \, , \quad r_1 = \frac{m_1}{m_2} r_2 \, , \quad \theta_1 = \theta_2 \, , \quad \phi_1 = \phi_2 \, .$$

Einstein applied this result in an asymptotic sense to the mass of asymptotically Schwarzschild space-times to obtain the variational result

$$\frac{\delta m}{\delta g_{\mu\nu}} = 0 \quad \Rightarrow \quad m = 0 .$$

The notion of asymptotic flatness used by Einstein was too over-simplified to include radiative space-times, and the thesis embodied in the title of his paper is generally conceded to be wrong. However, the energy scaling property which Einstein used has been established in a rigorous geometrical sense by D. Robinson and myself [2] for the most general asymptotically flat space-time by using Penrose's [3] description of infinity in terms of conformal geometry. Penrose's techniques lead to a geometrically meaningful formulation of the rest energy $E(\Sigma_+)$ of asymptotically flat systems as a scalar functional of retarded times Σ_+ [4],[5],[6]. The retarded times are geometrically described by outgoing null hypersurfaces extending to null infinity. The details of this formalism are not essential for the present considerations. I only wish to stress that the physical concepts of asymptotic flatness and energy have been geometricized.

It is clear that scale transformations must preserve asymptotic flatness. Thus it is possible to compare the energies of the one parameter set of asymptotically flat systems with metrics $\lambda^2 g_{\mu\nu}$ lying along the same ray. The result is that their energies satisfy the scaling law [2]

$$E_\lambda(\Sigma_+) = \lambda E_1(\Sigma_+).$$

In Sec. IV, it will be shown that this scale homogeneity of energy has a bearing on the sign of the energy. Before doing this, it is necessary to introduce some results which apply to linearized gravitational theory.

III. The Question of Positive Energy

Energy in general relativity refers to the active gravitational mass of a system in the same sense as the mass which appears in the Schwarzschild solution. From this viewpoint, the ground state of the hydrogen atom is not bound energetically because, although the sum of kinetic energy and potential energy is negative, the total energy (including rest masses) is positive. The hydrogen atom is only quantum mechanically bound due to restrictions against using its rest energy. A physically crucial question is: Can an absolutely bound state of negative rest energy arise, say, from a highly relativistic binary star system which eventually loses more energy by gravitational radiation than it began with? For such a situation to be physically interesting the matter sources representing the stars must be non-singular and have an energy momentum tensor corresponding to positive energy density. This rules out unrealistic solutions such as the Schwarzschild solution with negative mass. The important question is whether the gravitational contribution to the total energy can give rise to negative rest energy. This possibility is best isolated by investigating vacuum fields. Can, say, the scattering of two gravitational waves leave a negative energy remnant?

The history of significant research on this question is not very old. This reflects the mathematical complexity of the problem. The first results for weak gravitational fields were obtained by Araki [7] (1959) and later improved by Brill and Deser [8] (1968). Their approach assumes the existence of a space-like hypersurface extending to spatial infinity in whose neighborhood space-time differs from flat space to first order in some parameter ϵ. They then show that the energy measured at spatial infinity is of order ϵ and positive-definite to that order. This result implies that flat space is a strict local minimum of the energy measured at spatial infinity with respect to variations of geometry consistent with Einstein's equations.

The energy measured at spatial infinity corresponds in a limiting sense to the energy $E(\Sigma_+)$ measured at null infinity for infinitely past retarded times This limit is well-defined in terms of Penrose's conformal treatment of infinity, although there are certain singular aspects to the conformal properties of spatial infinity. Whereas the energy measured at spatial infinity is a constant, Bondi, van der Burg, and Metzner [9] (1962) have shown that for radiating systems $E(\Sigma_+)$ is a strictly decreasing functional of retarded time Σ_+. This positive-definiteness of the rate of energy loss suggests that a radiative system with initially positive energy may end up with negative energy at a later retarded time. Results by D. Robinson and myself [10] show that this cannot happen to lowest order in perturbations from flat space. At arbitrary retarded times the energy $E(\Sigma_+)$ has a strict local minimum at flat space with respect to variations of geometry. More precisely, consider a one-parameter family of solutions of Einstein's equations $g_{\mu\nu}(x; \epsilon)$ such that the curvature tensor satisfies

$$R_{\mu\nu\rho\sigma}(x;0) = 0, \quad \frac{d}{d\varepsilon} R_{\mu\nu\rho\sigma}(x;0) \neq 0$$

in a neighborhood of the null hypersurface corresponding to retarded time Σ_+.
Then $E(\Sigma_+; \varepsilon)$ satisfies

$$E(\Sigma_+;0) = \frac{d}{d\varepsilon} E(\Sigma_+;0) = 0, \quad \frac{d^2}{d\varepsilon^2} E(\Sigma_+;0) > 0.$$

The above results show that systems to which linearized gravitational theory
is applicable have positive energy. Also, Brill [11] (1959) has shown that
the energy measured at spatial infinity is positive for systems possessing a
moment of time symmetry. However, in the general case the question of the sign
of energy is still unresolved.

The historical importance of positive energy in physics coupled with these
partial results seems to have somewhat prejudiced research on this issue. There
are many incorrect proofs in the literature that the energy must be positive and
few incorrect proofs that it can be negative. One recent line of attack due to
Brill, Deser, and Fadeev [8], [12], which initially looked promising, is based
upon the variational result

$$\frac{\delta E}{\delta g_{\mu\nu}} = 0 \Rightarrow E = 0.$$

They first note from the weak field results that the energy has a strict local
minimum at flat space. They then couple this with the lack of other critical
points in the function space of space-times to argue that the energy must remain
positive. However, far from applying to the infinite-dimensional case of function
space, Geroch has pointed out that this argument holds only in one-dimension and
already breaks down in the two-dimensional plane. Geroch's technique of
constructing counterexamples is quite simple. Take any function with a strict
local minimum of zero, negative regions, and a finite number of other critical
points. Then use the property that any simply-connected portion of the plane
is diffeomorphic to the entire plane to cut out the extra critical points while

leaving the local minimum and at least part of the negative regions intact.

IV. Energy and the Neighborhood of Flat Space

An argument for positive energy which would be valid if the space of space-times were finite-dimensional can be based upon the homogeneous dependence of energy on scale transformations [2] . The chief usefulness of this argument is that it is quite simple to see the extra properties necessary for its extension to function space.

Let me first present this argument by assuming that the energy functional acts on a n-dimensional space. In that case the strict local minimum F (corresponding to flat space), with $E(F) = 0$, can be surrounded with a small (n-1)-dimensional sphere, S, on which the energy is positive. Suppose that these were a geometry G such that $E(G) < 0$. For physically interesting cases, there will be a path of geometries connecting G to F along which the energy is finite. The maximum energy attained along such a path is bounded from below by the positive, non-zero minimum energy on S. Hence, the greatest lower bound m of the maxima on all possible paths from G to F satisfies $m > 0$. While there need not be a path for which the greatest lower bound m is actually attained, for arbitrarily small ε there will exist a path with maximum $(1 + \varepsilon)m$. Given such a path, we now use the scale homogeneity to construct a new path in the following way. Since $E(G) < 0$ and $E(F) = 0$ the maximum $(1 + \varepsilon)m$ must be an interior point of the path. We can surround it with points G_1 and G_2 whose energy is $(1 - \varepsilon)m$ such that $(1 - \varepsilon)m$ is the maximum energy attained in the end segments $[F, G_1]$ and $[G_2 , G]$. We construct the new path from F to G by scale transforming the interior segment with scale factor λ ranging from 1 to ½. The deformed path is illustrated in Fig. 1. It has maximum energy $(1 - \varepsilon)m$. But m was the greatest lower bound on the maxima, so that we have reached a contradiction to the negative energy assumption $E(G) < 0$.

The above argument shows that Geroch's method of constructing counter-examples described in Sec. III does not apply to finite-dimensional homogeneous functions (of non-zero homogeneity order). If the function space of space-times

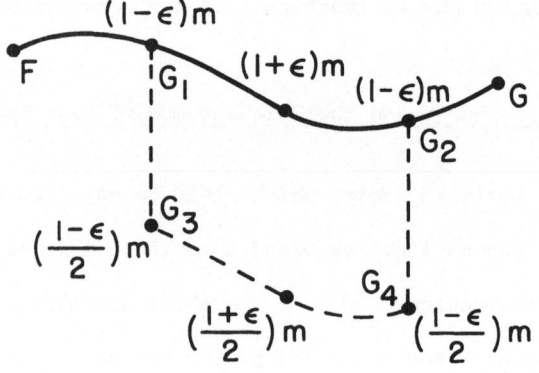

FIGURE 1

The deformed path is obtained by replacing the segment $[G_1, G_2]$ by the segments $[G_1, G_3]$ along which λ varies from 1 to $\frac{1}{2}$, $[G_3, G_4]$ along which $\lambda = \frac{1}{2}$, and $[G_4, G_2]$ along which λ varies from $\frac{1}{2}$ to 1.

were indeed finite-dimensional we could also conclude that negative energy is incompatible with the properties of a non-singular asymptotically flat space-time. However, there is one aspect of the scaling argument which depends critically upon finite-dimensional analysis and cannot be extended in any straightforward way. In the function space case, even if the strict local minimum property of flat space implied the existence of a small Jordan surface surrounding the minimum on which the energy were positive, in function space such a surface is non-compact. The energy need not attain a minimum on the surface. Consequently, in the foregoing argument, the greatest lower bound of the maxima taken over all paths from G to F might be zero. This possibility necessitates replacing the condition m > 0 by m 0. The added possibility m = 0 clearly bars extension of the argument.

This obstacle is not entirely due to the lack of local compactness of function space, but due to the special nature of the function space of space-times. Consider the case of a homogeneous functional E(f) on the space of real functions {f}, so that, say, E(λf) = λE(f). If such a functional had a strict local minimum it would have to be at f = 0. We could then trace back the function f along the monotonic scale path f_λ = λf until we reached the minimum at λ = 0 to show that E(f) is positive.

Unfortunately - or for negativists fortunately -, this one parameter scale path cannot be so used in the space of space-times. Consider the ray of geometries ds_λ^2 = $\lambda^2 ds^2$. In the limit $\lambda \to 0$, distances between corresponding points shrink to zero and the curvature becomes concentrated into a singularity. This can be seen from the scale behavior of a typical curvature invariant

$$I = R_{\mu\nu\rho\sigma} R^{\mu\nu\rho\sigma} ;$$

in the limit $\lambda \to 0$, it is clear that

$$I_\lambda = \lambda^{-4} I \to \infty .$$

A certain amount of caution must be exercised here. Geroch [13] has shown that the limit of a 1-parameter family of space-times is quite ambiguous.

He defines a 1-parameter family of space-times as a 5-dimensional manifold with (degenerate) metric g^{ab} of signature $(0 + - - -)$ and scalar field τ . The 4-dimensional space-times M_τ are given by the hypersurfaces τ = constant. The various limits of the one-parameter family M_τ correspond to permissible boundaries on this 5-manifold. The ambiguity in the limit results from the action of a 1-parameter family of diffeomorphisms of the intermediate space-times M_τ .

According to this formalism, one possible limit of the Schwarzschild geometry is Minkowski space with no points missing [13] . This is achieved by shoving the source farther away at each stage so that it is at infinity in the limit. More explicitly, applying the coordinate transformation

$$z = r - \tau^{-1} \quad , \quad \rho = \tau^{-1} \theta \ ,$$

to the conventional form of the Schwarzschild metric (as given in Sec. II) leads to the 1-parameter family M_τ with metrics

$$ds_\tau^2 = \left(1 - \frac{2\tau m}{1 + \tau z}\right) dt^2 - \left(1 - \frac{2\tau m}{1 + \tau z}\right) dz^2$$

$$- (1 + \tau z)^2 \left(d\rho^2 + \tau^{-2} \sin^2 \tau \rho \ d\phi^2\right) \ ,$$

and coordinate ranges

$$-\infty < t < +\infty, \quad -\tau^{-1} < z < \infty, \quad 0 \leq \rho \leq \tau^{-1}\pi \ , \quad 0 \leq \phi \leq 2\pi.$$

The limit $\tau \to 0$ directly gives the entire Minkowski geometry in cylindrical coordinates. This is in spite of the fact that each intermediate space-time M_τ is isometric to the original Schwarzschild geometry. Furthermore, the rest energy, being a scalar functional, is independent of τ . Consequently, the limit of the rest energy is the initial energy while the limit of the geometry is flat space. The same flat space limit of Schwarzschild space can also be obtained while varying the mass along the scale path $m \to 0$ or $m \to \infty$.

These limits seem to contradict the weak field results that the energy approaches zero as the geometry approaches flat space. The explanation is that although each intermediate space-time is asymptotically flat, the approach to flat space is through a non-asymptotically flat direction. In other words, the linearized solution induced by such a path is not asymptotically flat. This can be seen for the above limit of Schwarzschild space by examining the only independent non-vanishing component of the Riemann tensor,

$$R_{tztz} = \frac{2m}{\left(z + \tau^{-1}\right)^3} \sim 2m\tau^3 .$$

The extremal property of the energy at flat space applies only to asymptotically flat perturbations. It is plausible that limits of a 1-parameter family of asymptotically flat space-times M_τ which do not induce asymptotically flat linearized solutions could be ruled out by considering the M_τ to be manifolds with boundary as prescribed by Penrose's conformal techniques. However, whether such a scheme could be completely successful is not readily apparent due to certain complications [3] regarding the conformal structure of space-like and time-like infinities.

The importance of the scaling argument is that it reduces the question of positive energy to considerations of a neighborhood of flat space. If flat space can be surrounded by a small Jordan surface on which the energy is positive and bounded away from zero, then the energy must be positive. A better understanding of what topological structure can be given to the space of space-times is clearly needed to decide this matter. The main part of the scaling argument involves topological properties only of 1-parameter paths of space-times, as provided by Geroch's work. It is the missing link of supplying the Jordan surface in the neighborhood of flat space in which topological properties of the space of space-times must be introduced.

The important physical problem of the sign of gravitational energy thus leads to a fundamental question: How are properties of the linearized theory to be correctly formulated as properties of the neighborhood of flat space? The physicist is often content with the intuitive answer to this question that linearized theory is a good approximation to weak gravitational fields. But the issues of this paper emphasize that there is a need for a proper mathematical understanding of the concept of weak gravitational fields.

REFERENCES

1. A. Einstein, Universidad Nacional de Tucuman Revista A2, 11, (1941).

2. D. C. Robinson and J. Winicour, "Scaling Behavior of Gravitational Energy", preprint.

3. R. Penrose, Proc. Roy. Soc. (London) A284, 159 (1965).

4. R. Penrose in Relativity and Astrophysics, ed. J. Ehlers (American Mathematical Society, 1967).

5. L. Tamburino and J. Winicour, Phys. Rev. 150, 1039 (1966).

6. J. Winicour, J. Math. Phys. 9, 861 (1968).

7. H. Araki, Ann. Phys. (N.Y.) 7, 456 (1959).

8. D. Brill and S. Deser, Ann. Phys. 50, 548 (1968).

9. H. Bondi, M. G. I. van der Berg, and A. W. K. Metzner, Proc. Roy. Soc. (London) A270, 103 (1962).

10. D. C. Robinson and J. Winicour, "Can the Energy of a Weak Gravitational Field Become Negative?", preprint.

11. D. Brill, Ann. Phys. (N.Y.) 7, 466 (1959).

12. D. Brill, S. Deser and C. Fadeev, Phys. Letters 26A, 538 (1968).

13. R. Geroch, Commun. Math. Phys. 13, 180 (1969).

ON THE SPHERICAL SYMMETRY

OF A STATIC PERFECT FLUID*

H.P. Künzle**

University of California

Berkeley

1. Introduction: Newtonian hydrostatics

The equilibrium configurations of a perfect fluid subject to its own Newtonian gravitation were studied extensively at the beginning of this century, and the results are summarized in the classical book of LICHTENSTEIN [4]. In particular, the intuitively expected result that a body consisting of a perfect fluid at rest in empty space will have a spherical shape was shown to be a direct consequence of the governing equations and a natural boundary condition.

The assumptions are the following:

(a)   the underlying manifold Σ is diffeomorphic to Euclidean space : $\Sigma \cong \mathbb{R}^3$,

(b)   the metric γ on Σ is the standard Euclidean metric,

(c)   the potential U and the mass density satisfy Poisson's equation:

$$\Delta U = 4\pi\rho,$$

(d)   the equation of hydrostatic equilibrium holds between pressure, density and the potential: $dp + \rho \, dU = 0$,

(e)   the fluid body is finite, i.e. supp ρ and supp p are compact,

(f)   for $r = |x| = \sqrt{\sum_{i=1}^{3} (x^i)^2} \longrightarrow \infty$, $U = 0(r^{-1})$ and $\partial_i U = 0(r^{-2})$,

(g)   all functions are sufficiently regular; e.g., U is C^1 and piecewise C^2,

*   This work was supported in part by the United States Atomic Energy Commission under contract number AT(04-3)-34, Project Agreement No.125.

** Present address: Department of Mathematics, University of Alberta, Edmonton 7, Canada .

(h) the function ρ is nowhere negative.

These assumptions imply that U, p and ρ are actually functions of the radial distance r only (if the origin is suitably chosen) and thus that the whole system is spherically symmetric. Moreover, the function U has only one critical point, namely the center r=0, which is a nondegenerate minimum.

Physically it seems clear that with some suitable modifications this result should remain true for relativistic fluids.

2. Static perfect fluids in General Relativity

The equations of the static relativistic perfect fluid can be put into a very similar form. Following LICHNEROWICZ [4] we call a spacetime M (globally) static if there exists a three-dimensional manifold Σ and a diffeomorphism ψ : M → Σ × ℝ such that

(i) $C_x = \psi^{-1}(\{x\} \times \mathbb{R})$ are timelike curves for all x ε Σ,

(ii) $\Sigma_t = \psi^{-1}(\Sigma \times \{t\})$ are (globally) spacelike hypersurfaces for all t ε ℝ,

(iii) C_x for all x ε Σ are tangent to a Killing vector field ξ on M that is orthogonal to all Σ_t.

We assume also that M is geodesically complete and orientable; then so is Σ with its induced metric. The metric on M can be characterized by

$$g = \psi^*(-e^{2U}dt^2 + e^{-2U}\gamma) \tag{1}$$

where γ is a complete Riemannian metric and U a real-valued function on Σ and dt^2 the standard metric on ℝ. The function U is called the gravitational scalar potential of the static field. The space-time or the corresponding (Σ,γ,U) is called spherically symmetric if SO(3) is an isometry group of (M,g) whose orbits are (generically) spacelike hypersurfaces.

If Einstein's equations for a perfect fluid

$$\overset{4}{R}_{\alpha\beta} - \frac{1}{2}\overset{4}{R}g_{\alpha\beta} = 8\pi\left[(\rho+p)u_\alpha u_\beta - pg_{\alpha\beta}\right] \tag{2}$$

are written out for the metric (1) it becomes clear that the conditions
(a) to (h) of section 1 must be modified as follows:

(a') Σ is an orientable three-dimensional manifold;

(b') γ is a complete Riemannian metric on Σ;

(c') U, γ, ρ and p are related by

$$\Delta U = 4\pi(\rho+3p)e^{-2U} \qquad \text{and} \qquad (3)$$

(d') $$R_{ij} = 2\partial_i U \partial_j U - 16\pi p e^{-2U}\gamma_{ij} , \qquad (4)$$

where all geometric quantities refer to γ. The Bianchi identities for R_{ij} are
then equivalent to the conditions that (i) the fluid satisfies an equation of
state of the form $\rho = \rho(p)$ and (ii) pressure and density are functions of U only
and are determined up to a constant by the equation of state by means of

$$dp + (\rho+p)dU = 0, \qquad (5)$$

the analogue of the equation of hydrostatic equilibrium. We take over condition
(e) unchanged and replace (f) by

(f') (Σ, γ, U) is <u>asymptotically Euclidean</u>, i.e.

(i) there exists a compact $K \subseteq \Sigma$ and a diffeomorphism
     $\phi: \Sigma - K \rightarrow \mathbb{R}^3 - B$, where B is a closed ball in $\mathbb{R}^3$ centered at the
     origin,

(ii) with respect to the standard coordinate system in $\mathbb{R}^3$ for

$$|x| \rightarrow \infty : \quad \gamma_{ij} - \delta_{ij} = 0(|x|^{-1}) = U \qquad \text{and}$$

$$\partial_k \gamma_{ij} = 0(|x|^{-2}) = \partial_k U.$$

Instead of (g) we assume the Lichnerowicz conditions:

(g') γ and U are C^1 and piecewise C^2 with respect to the differentiable structure
of Σ (which is at least C^2).

Because also pressures are sources of gravitational energy in general relativity the condition $\rho \gtrsim 0$ is not enough to imply the desired result in the relativistic case and some restrictions on the pressure are needed. But a necessary and sufficient condition seems to be unknown. We assume for simplicity the certainly physically reasonable but probably unnecessarily restrictive conditions

(h')
$$0 \le 3p \le \rho \qquad \text{and} \qquad 0 < \frac{dp}{d\rho} = \alpha^2 \le \infty$$

where α is interpreted as the velocity of sound in the fluid.

We summarize some simple consequences of these assumptions (see LICHNEROWICZ [4] and EHLERS and KUNDT [2]) in

THEOREM 1:

(a) $U < 0$ on Σ,

(b) U has no maximum and no minimum in the vacuum region and no critical point in some neighborhood of ∞ unless the space is flat,

(c) $D_c = \{x \in \Sigma \ / \ U(x) \le c\}$ is compact for all $c < 0$,

(d) $S_c = \partial D_c = U^{-1}(\{c\}) \cong S^2$ for c sufficiently small,

(e) $U = -\frac{m}{|x|} + 0(|x|^{-2})$,
$W = |\nabla U| = (\gamma^{ij}\partial_i U \partial_j U)^{\frac{1}{2}} = \frac{m}{|x|^2} + 0(|x|^{-3})$
for $|x| \to \infty$, where m is a constant in vacuum, in fact, the total (active gravitational) mass of the system, defined by

$$m = \int_D (\rho+3p) e^{-2U} \sqrt{\gamma} \ d^3x \ ,$$

where D is any domain of Σ containing supp($\rho+3p$).

3. Results

AVEZ [1] proved by an elegant argument based on Morse theory that if the gravitational field strength W is a function of the potential U only and U has no degenerate critical points on Σ, then Σ is diffeomorphic to $\mathbb{R}^3$, (Σ,γ,U) is spherically symmetric and U has only one critical point, a nondegenerate minimum at the center. This is the result one would expect to follow from conditions (a') to (h') alone, without additional assumptions.

While a general proof seems still to be unknown two partial improvements
have been achieved, the first to drop the condition that U be a Morse function in
Avez's result, the second the proof of an infinitesimal version of the general
theorem.

THEOREM 2:

If, for a static space-time (M,g) representing a finite body of a perfect
fluid with $\rho+3p > 0$ and an asymptotically Euclidean vacuum region, the
gravitational field strength W is a function of U then M is diffeomorphic to
$\mathbf{R^4}$ and spherically symmetric. The potential U is a monotonically increasing
function of the radial distance from the center.

In view of the special assumption made this result is of limited interest in its
own right, but there is some hope that $W = W(U)$ can be deduced directly from the
equations of a static perfect fluid under the assumption of asymptotical flatness.

By an infinitesimal version of the general theorem on spherical symmetry is
meant the following: Let $\mathfrak{S}$ be the set $\{(\gamma,U)\}$ of asymptotically Euclidean pairs
of metrics γ and potentials U on $\Sigma \cong \mathbf{R^3}$ that satisfy the equations of a perfect fluid
with a given equation of state $\rho = \rho(p)$ and a fixed $U_0 = \min (U/x\epsilon\Sigma)$ and $p_0=p(U_0)$.
There is a (up to isometry) unique $(\tilde{\gamma},\tilde{U})\epsilon\ \zeta$ that is spherically symmetric and, in
any case, p and ρ are uniquely determined as functions of U by the equation of
state and (5) and the given initial conditions. The group Diff(Σ) of all
diffeomorphisms of Σ acts on ζ in a natural way:

$$\psi:\ \text{Diff}(\Sigma) \times \mathfrak{S} \longrightarrow \mathfrak{S}$$
$$(a\ ,\ (\gamma,U)) \qquad (\psi_a^*\gamma, \psi_a^* U)$$

If $\psi_{(\gamma,U)}(a)= \psi(a,(\gamma,U))$ for all $a\ \epsilon$ Diff(Σ) then $\psi_{(\gamma,U)}$(Diff(Σ)) is the orbit of
(γ,U) under Diff(Σ).

Our aim is to show that the tangent space to the orbit through $(\tilde{\gamma},\tilde{U})$ coincides
with the whole tangent space of ζ at $(\tilde{\gamma},\tilde{U})$, or that

$$T_e\psi(\gamma,U)\ :\ T_e(\text{Diff}(\Sigma)) \rightarrow T_{(\gamma,U)}\ \zeta$$

(e = identity in Diff(Σ)) is surjective. Then $(\tilde{\gamma},\tilde{U})$ is called infinitesimally stable under Diff(Σ). In other words, any first order deformation of the spherically symmetric solution in ζ arises necessarily from a mere coordinate change.

THEOREM 3:

A static spherically symmetric asymptotically Euclidean space-time representing vacuum and a perfect fluid with $0 \le 3p \le \rho$ and an equation of state $\rho = \rho(p)$ satisfying $0 < \dfrac{dp}{d\rho} \le \infty$ in a compact domain does not admit any (nontrivial) static deformations with the same equation of state, the same maximal pressure and the same minimal value of the gravitational potential.

Remark: The minimal value of U is fixed in the definition of ζ for technical reasons. It is probably equivalent to fixing the total mass m.

4. Method of proof *)

The fact that the equipotential surfaces S_c are compact for $c < 0$ is used throughout in an essential way. If

$$\gamma = \gamma_{11} dU^2 + 2P_A dx^A dU + \bar{\gamma}_{AB} dx^A dx^B \quad (A,B = 2,3) \quad (6)$$

where $\bar{\gamma}_{AB}$ is the induced metric on S_c and if $\Omega_{AB} = \nabla_A(W^{-1}\partial_B U)$ is the second fundamental form, then it follows from (3) and (4) that

$$\partial_U W - P^A \partial_A W = 4\pi(\rho+3p)W^{-1}e^{-2U} -\Omega, \quad (7)$$

$$W(\partial_U \Omega - P^A \partial_A \Omega) - W^{-1}\overline{\Delta W} + 2W^{-2}\bar{\partial}^A W \partial_A W + \Omega_{AB}\Omega^{AB}$$

$$- 2W^2 - 16\pi p e^{-2U} = 0, \quad (8)$$

$$\partial_A \Omega = \bar{\nabla}_B \Omega_A^B \quad \text{and} \quad (9)$$

*) A paper containing complete proofs has been submitted to the Commun. math. Phys.

$$\bar{R} \, - \, \Omega^2 \, + \, \Omega_{AB}\Omega^{AB} \, + \, 2W^2 \, + \, 16\pi p e^{-2U} \, = \, 0, \tag{10}$$

where all barred quantities refer to $\bar{\gamma}_{AB}$ and $\Omega = \bar{\gamma}^{AB}\Omega_{AB}$.

If now $W = W(U)$, then $\partial_A W = 0$ implies by (7) that $\Omega = \Omega(U)$, then by (8) that $\Omega_{AB}\Omega^{AB}$ and by (10) that $\bar{R}$ is a function of U only. Thus S_c is of constant curvature. Applying the

Lemma: "If M is a two-dimensional Riemannian manifold and K_{AB} a symmetric tensor field such that $\bar{\nabla}^A K_{AB} = 0$ and $K = \bar{\gamma}^{AB} K_{AB} =$ const. then there exists a harmonic 1-form ϕ on M such that $K_{AB} = \frac{1}{2}(K - \phi_C\phi^C)\bar{\gamma}_{AB} + \phi_A\phi_B$."
to equation (9) shows that $\Omega_{AB} = \frac{1}{2}\Omega\bar{\gamma}_{AB}$ whence it follows that the metric $\bar{\gamma}$ is spherically symmetric in any domain where U has no critical point. Equations (7) and (8) reduce then to two ordinary first order differential equations for W and the radius of curvature r of S_c as functions of U which can easily be discussed. A local consideration of a neighborhood of a critical point x_o of U moreover reveals that x_o is necessarily a nondegenerate minimum, and the original argument of MORSE [5] then leads to theorem 2.

For the proof of theorem 3 we observe that the property of U of having only one nondegenerate critical point in the spherically symmetric case is stable under small static deformations, so that U can be used as a coordinate also for solutions near the spherically symmetric one. We consider therefore the in-finitesimal deformations h_{AB}, ωW and Q_A of the quantities $\bar{\gamma}_{AB}$, W and P_A satisfying the linearized versions of equations (7) to (10). But on the compact Riemannian manifold S_c, h_{AB} can be decomposed into a uniquely defined symmetric tensor ϕ_{AB} with vanishing divergence and a Killing derivative,

$$h_{AB} \, = \, \phi_{AB} \, + \, 2\bar{\nabla}_{(A}\xi_{B)}$$

with $\bar{\nabla}^A\phi_{AB} = 0$, where ξ_A is given up to an arbitrary Killing vector field of $\bar{\gamma}_{AB}$. From the linearized equations (7) to (10) it follows then together with the conditions of asymptotical flatness and regularity at the center that $\omega = 0 = \phi_{AB}$

and $Q^A = \partial_U(\xi^A)$. This, however, means that the deformation of (γ, U) is due to an infinitesimal coordinate transformation.

For this part of the proof the conditions $3p \leq 0$ and $\dfrac{dp}{d\rho} > 0$ were needed.

The author is grateful to Professors A.Avez and A.Taub for several helpful discussions.

References

1 A.AVEZ, Ann. Inst. Henri Poincaré <u>1</u>, 291-300, 1964.

2 J.EHLERS and W.KUNDT, article in L.WITTEN (ed.) <u>Gravitation, an introduction to current research</u>, Wiley, New York, 1962.

3 A.LICHNEROWICZ, <u>Théories relativistes de la gravitation et de l'éléctromagnetisme</u>, Masson, Paris, 1955.

4 L.LICHTENSTEIN, <u>Gleichgewichtsfiguren rotierender Flüssigkeiten,</u> Springer, Berlin, 1933.

5 M.MORSE, Trans. Amer. Math. Soc. <u>27</u>, 345-396, 1925.

DIFFERENTIABLE MANIFOLDS WITH SINGULARITIES

R. Reynolds

Pennsylvania State University at McKeesport

There seems to be general agreement here that the problem of singularities is a most fascinating one. Furthermore, the primary difficulty is that of arriving at a reasonable, and at the same time workable, definition of exactly what singularities are. This situation is not peculiar to relativity theory. What I want to present is a similar problem and its solution in the area of differentiable manifolds.

At a first glance, the term differentiable manifold with singularities seems contradictory, since by definition, a differentiable manifold does not have singular points; that is, its definition does not single out one point having properties different from the others. However, I think anyone who has studied differentiable manifolds has some intuitive concept of a topological space which would be a differentiable manifold except for certain points of the space at which it cannot have a manifold structure. For example, a curve or a surface in three space which intersects itself would be a manifold except for the points of intersection. Of course such a space would be handled classically as an immersion of a manifold. The simplest example which is not an immersion is the cone in $\mathbb{R}^3$ with equation $x^2 + y^2 = z^2$, which is a manifold except at its vertex. A slightly more complicated example is the cone in three space with its vertex at the origin and a figure eight curve for its directrix.

In this talk I shall display the essential features of the theory of manifolds with singularities primarily using the cone as an example. The analysis of the cone $x^2 + y^2 = z^2$ which will lead to an appropriate definition consists of two steps. The first is to realize that, although the cone does not have a coordinate chart at the vertex, it has the next best thing, namely a cylindrical coordinate parameterization $(\theta, z) \rightarrow (z \cos \theta, z \sin \theta, z)$, which would be a coordinate chart except that it collapses to a point on $z = 0$. The second step is to realize that this parameterization can be factored by the cylindrical coordinate parameterization $(\theta, z) \rightarrow (\cos \theta, \sin \theta, z)$ of the cylinder $x^2 + y^2 = 1$ in $\mathbb{R}^3$. Thus $\big((\theta, z) \rightarrow (z \cos \theta, z \sin \theta, z)\big) = \big((\theta, z) \rightarrow (\cos \theta, \sin \theta, z)$

$$\rightarrow (z \cos \theta, z \sin \theta, z)\big).$$

By suitably restricting θ this parameterization gives a set of coordinate charts for the cylinder (Figure 1).

Definition. Let M be a topological space and R an open subspace. Let $S = M-R$. The pair (M, S) is said to be an n-dimensional C^r manifold with singularities S and with boundary if there is a collection $\{(\phi_i, U_i)\}$ of open subsets U_i of H^n and continuous maps $\phi_i \rightarrow M$ such that:

(1) $\bigcup_i \phi_i(U_i) = M$.

(2) $\phi_i^{-1}(R)$ is dense in U_i for each i.

(3) the restriction of each ϕ_i to R is one to one and the topology on M is the final topology induced by $\{(\phi_i, U_i)\}$. That is, a subset A of M is open if and only if each $\phi_i^{-1}(A)$ is open in U_i.

(4) if int $\phi_i^{-1}(\phi_i(U_i) \cap \phi_j(U_j))$ is nonempty for some i and j then there is a (necessarily unique) C^r one to one map

h_{ji}:int $\phi_i^{-1}(\phi_i(U_i) \cap \phi_j(U_j)) \rightarrow$ int $\phi_j^{-1}(\phi_i(U_i) \cap \phi_j(U_j))$ such that $\phi_j h_{ji} = \phi_i$ on int $\phi_i^{-1}(\phi_i(U_i) \cap \phi_j(U_j))$.

We construct a manifold M by identifying each $h_{ji}(p)$ with p. It is easy to show that this is a C^r n-manifold in the usual sense and that there is a map $\sigma: \overline{M} \rightarrow M$ such that each $\sigma_i = \sigma \psi_i$ where each ψ_i is a coordinate chart for $\overline{M}$. The manifold $\overline{M}$ is called the singular covering manifold for M. It will be required that σ be one to one when restricted to the regular points of M.

We now use the map $\sigma: \overline{M} \rightarrow M$ to transfer the differential structure of $\overline{M}$ from $\overline{M}$ down onto M.

If $\sigma_M: \overline{M} \rightarrow M$ and $\sigma_N: \overline{N} \rightarrow N$ are singular manifolds with their singular covers, then a map $f: A \rightarrow N$ on an open set in M is called singular differentiable if there is a differentiable map $\overline{f}: \overline{M} \rightarrow \overline{N}$ such that

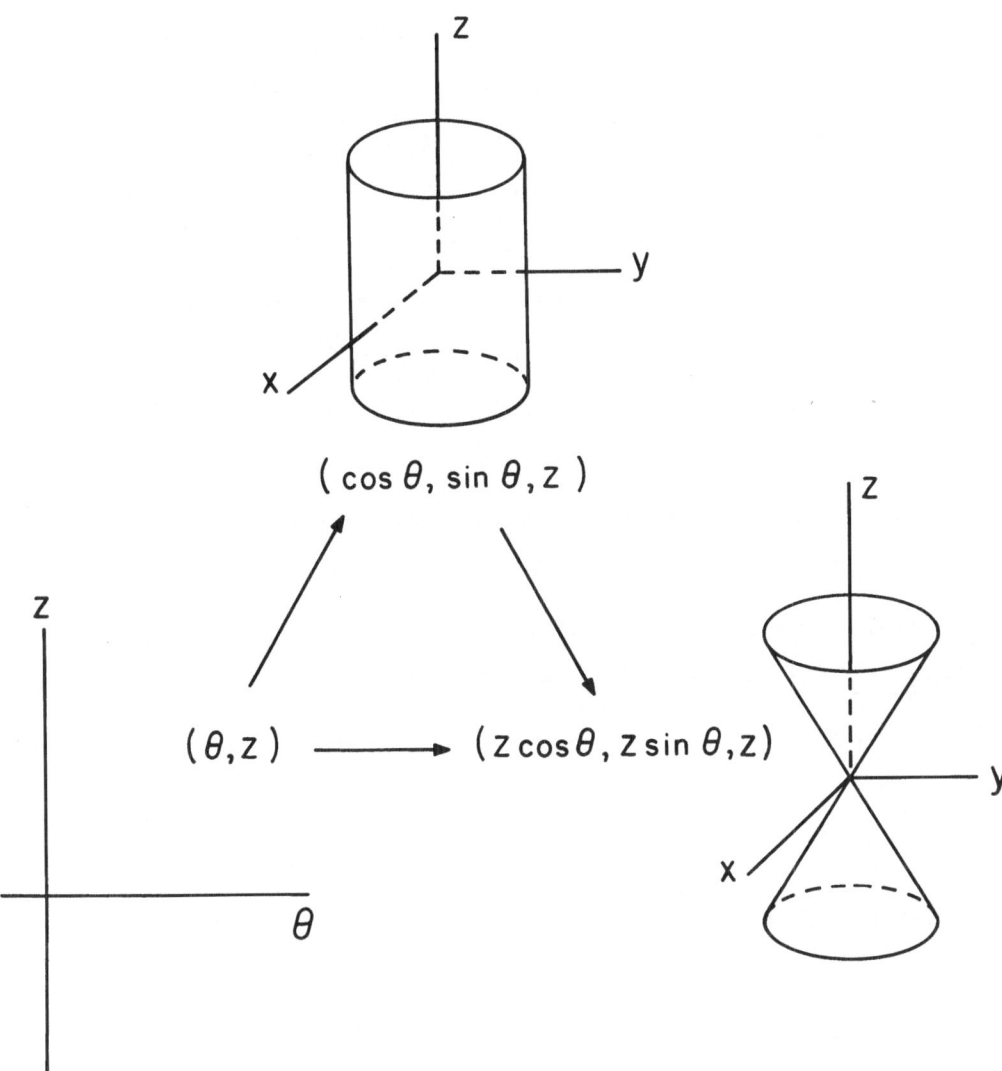

Figure 1

$\sigma_N \overline{f} = f\sigma_M$. In particular a real valued function is singular differentiable if $f\sigma$ is differentiable and a curve $\alpha: [a,b] \to M$ is singular differentiable if there is a differentiable curve $\overline{\alpha}: [a,b] \to \overline{M}$ such that $\alpha = \sigma\overline{\alpha}$.

If we let $\mathcal{F}_p$ denote the set of a real valued functions which are singular differentiable on some neighborhood of a point p in M and if α is a singular differentiable curve with $\alpha(0) = p$ then the tangent vector $\alpha'(0)$ at p is defined as the map $\alpha'(0): \mathcal{F}_p \to \mathbb{R}$ where $\alpha'(0)(f) = (f\alpha)'(0)$.

The set M_p of all tangent vectors at a point p of M is called the tangent cone at that point. In general we cannot add two tangent vectors at a singular point to get another tangent vector. However, we can scalar multiply a tangent vector and get another tangent vector. The scalar multiplication obeys the axioms of the following algebraic definition of a cone.

<u>Definition</u>. Let $\mathbb{R}$ denote the real numbers. A <u>cone</u> K over $\mathbb{R}$ is a set K together with a map

$$\mathbb{R} \times K \to K$$

$$(a,x) \to ax$$

which satisfies the following axioms:

(1) $a(bx) = (ab)x$ for every $a,b \in \mathbb{R}$ and $x \in K$.

(2) $1x = x$ for every $x \in K$.

(3) $0x = 0y$ for every $x,y \in K$.

The element $0x$ will be denoted by 0 and will be called the <u>vertex</u> of K.

(4) if $ax = x$ for some $x \neq 0$ then $a = 1$.

One can now proceed to study the category of cones and cone maps. I will not go into this here except to mention that the manner in which such objects as the direct sum, direct product, kernel, etc. are constructed is determined by general category theory, and frequently these constructions have geometric meaning. For example, the direct sum $\mathbb{R} \oplus \mathbb{R}$ of two copies of $\mathbb{R}$ is the cone consisting of the two lines intersecting at their origins.

Topological cones and cone bundles are also defined in a rather obvious manner.

The tensor algebras on a cone can now be developed in a very obvious manner,

and the tensor calculus on a singular manifold can be similarly developed. The
contravariant tensor calculus has one unpleasant aspect, namely that it is most
natural for a contravariant tensor field to have many values at a singular point.
For example a differentiable vector field on an open neighborhood of the vertex of
the cone $x^2 + y^2 = z^2$ must vanish at the vertex or else assume infinitely many
values there. On the other hand, the covariant tensor calculus generalizes easily.
Thus a covariant p-tensor field η on a Manifold M with singularities is a
function which assigned to each p-tuple $(X_1,\ldots,X_p)$ of vectors at a point x of M
a real number $\eta_x(X_1,\ldots,X_p)$ in such a manner that η_x is a cone map in each
variable $X_1,\ldots,X_p$. A p-form is an alternating p-tensor field. The concept of a
p-cell and the integral of a p-form over a p-cell generalize naturally. The only
disappointing feature of the differential forms on a manifold with singularities is
that there seems to be no natural definition of the exterior differential of a
p-form on a manifold with singularities, and therefore no DeRham cohomology.

At this stage it becomes more interesting to stop generalizing concepts from
the classical theory of manifolds, and to look for those properties which distinguish a singular point from a regular point. The most interesting property, which I
shall discuss here, is the size of a point.

First we define the type of a cone. The easiest way to do this precisely is to
use the concept of a cone bundle, but it can be described fairly well as the largest
integer k such that each point of the cone is contained in a linear space of
dimension k, which is a subcone of the cone. For example it is easy to see that
the cone $x^2 + y^2 = z^2$ in $\mathbb{R}^3$ is of type one. Of course the type of a cone may
not exist, as in the case of $\mathbb{R}^\infty$.

Let M be a manifold with singularities and p a point on M, with M_p the
tangent cone at p. We define the size of the point p by the equation, size(p) =
dim(M) – type(M_p). For example the vertex of the cone $x^2 + y^2 = z^2$ has size one
if the usual cylindrical coordinate parameterization is used. For then the cone is
its own tangent cone at the vertex. On the other hand if the parameterization
$(\theta,z) = (z^3\cos\theta, z^3\sin\theta, z^3)$ is used, then the tangent cone at the vertex is a

single point and is therefore type zero. Therefore the vertex is now size two. These examples show that the size of a point measures the amount which the singular covering manifold collapses onto the point, and also the amount that tangent spaces to the points covering the point collapse onto the tangent cone to the point.

Now what does this have to do with relativity theory? I do not think that this theory can be applied directly to the study of physical singularities. However, I think it is interesting to compare this approach to the problem with those approaches which were presented in an earlier talk. The approach so far to singularities in relativity has been integral in nature as opposed to this approach which is differential in nature. By this I mean that the other approaches have been to start with a manifold in the usual sense and a metric on that manifold and then attempt to attach the singular points to the manifold using the properties of the metric. This approach begins with the singularities already in the manifold and the knowledge of the topology of the manifold, and then studies the properties of the singular points. Now, I do not mean to suggest that the construction of a space-time manifold can be other than an integration process, but it might be helpful to study what the properties of a space-time manifold with singularities should be, from basic physical principles.

NON-VACUUM ADaM FIELD EQUATIONS*

Bernard F. Schutz, Jr. [†]

California Institute of Technology, Pasadena, California 91109

The canonical version of the vacuum Einstein field equations formulated ten years ago by Arnowitt, Deser, and Misner (ADaM) [1] has stimulated several attempts to quantize certain cosmological models, most notably Misner's so-called Mixmaster Universe [2] . Some researchers have begun recently to extend these methods to non-vacuum spacetimes; for example, Nutku earlier at this conference described the canonical theory of a scalar field in Schwarzschild spacetime. The purpose of this talk is to generalize the ADaM field equations to include an arbitrary stress-energy tensor. This is not a "first step" toward a canonical formulation of the full non-vacuum field equations; rather, it is simply a possible starting point.

Essentially, the ADaM field equations are a linear combination of Einstein's $G_{\mu\nu} = 0$ equations that is particularly well-suited to a "three-plus-one split" of spacetime, i.e., a division of spacetime into three-dimensional spacelike sections labelled by the parameter time. The metric of each section is the spacelike part of the metric for all of spacetime:

$$g_{ij} \equiv {}^4g_{ij} . \qquad (1a)$$

(Superscript "4" denotes quantities referred to the full four-dimensional spacetime, while no superscript implies three-dimensional quantities. Latin indices run from

---

*Supported in part by the National Science Foundation GP-15911, GP-9114, GP-19887 and the Office of Naval Research Nonr-220(47) .

[†]NDEA Title IV Pre-doctoral Fellow .

---

1 to 3, Greek from 0 to 3. Signature is − 2.) ADaM replace the remaining four metric components − which give information on how one hypersurface fits into the next 3 − with: a three-scalar

$$N \equiv (- \, {}^{4}g^{00})^{-\frac{1}{2}} \tag{1b}$$

and a covariant three-vector

$$N_i \equiv {}^{4}g_{0i} \, . \tag{1c}$$

The ADaM field equations are derived from the usual variational principle,

$$\delta I = \delta \int {}^{4}R(- \, {}^{4}g)^{\frac{1}{2}} \, d \, {}^{4}x = 0 \, . \tag{2}$$

Were one to use $\{{}^{4}g^{\mu\nu}\}$ as the set of independent variables, one would obtain $G_{\mu\nu} = 0$ from Eq. (2) [4]. Using the ADaM variables $\{N, N_i, g_{ij}\}$, on the other hand, gives the ADaM equations.

To obtain the non-vacuum equations, let L be the Lagrangian for the non-gravitational fields. Then Eq. (2) generalizes to

$$\delta I = \delta \int ({}^{4}R + 2 \, \kappa \, L)(- \, {}^{4}g)^{\frac{1}{2}} \, d \, {}^{4}x = 0 \, . \tag{3}$$

Using $\{{}^{4}g^{\mu\nu}\}$ as the variables gives [5]

$$G_{\mu\nu} = \kappa \, T_{\mu\nu} \, , \tag{4}$$

where

$$T_{\mu\nu} = L \, {}^{4}g_{\mu\nu} - 2 \, \frac{\partial L}{\partial \, {}^{4}g^{\mu\nu}} + \frac{2}{(- \, {}^{4}g)^{\frac{1}{2}}} \left[(- \, {}^{4}g)^{\frac{1}{2}} \, \frac{\partial L}{\partial \, {}^{4}g^{\mu\nu}{}_{,\beta}} \right]_{,\beta} \, . \tag{5}$$

The non-vacuum ADaM equations follow from Eq. (3) if one uses the set $\{a_{\alpha\beta}\}$ of ADaM variables, defined by

$$a_{00} \equiv (- \, {}^{4}g^{00})^{-\frac{1}{2}}; \; a_{0i} \equiv {}^{4}g_{0i}; \; a_{i0} \equiv {}^{4}g_{i0} \, ; \; a_{ij} \equiv {}^{4}g_{ij} \, . \tag{6}$$

It is convenient in what follows to ignore the symmetry of $a_{\alpha\beta}$ and ${}^{4}g_{\mu\nu}$. For instance, variations of a_{0i} will be taken while holding a_{i0} fixed. The final results will, of course, be symmetrized.

Because the transformation from $\left\{{}^4g^{\mu\nu}\right\}$ to $\left\{a_{\alpha\beta}\right\}$ is nonsingular and does not involve derivatives of ${}^4g^{\mu\nu}$ or explicit dependence upon the spacetime coordinates, the equations obtained from varying $a_{\alpha\beta}$ will be the linear combination

$$0 = \frac{\delta I}{\delta a_{\alpha\beta}} = \frac{\partial\, {}^4g^{\mu\nu}}{\partial a_{\alpha\beta}}\, \frac{\delta I}{\delta\, {}^4g^{\mu\nu}} \qquad (7)$$

of the equations obtained from varying ${}^4g^{\mu\nu}$. We therefore need only find $\partial\, {}^4g^{\mu\nu}/\partial a_{\alpha\beta}$, in which it is understood that the derivative is taken holding all other $a_{\gamma\delta}$ fixed. This is the key to the difference between Einstein and ADaM: it means, for example, that $\partial\, {}^4g^{01}/\partial a_{01}$ is not the same as $\partial\, {}^4g^{01}/\partial\, {}^4g_{01} = -{}^4g^{00}\, {}^4g^{11}$, because in the first case one holds $\left\{{}^4g^{00},\ {}^4g_{02},\ {}^4g_{03},\ {}^4g_{ij}\right\}$ fixed while in the second case one holds $\left\{{}^4g_{00},\ {}^4g_{02},\ {}^4g_{03},\ {}^4g_{ij}\right\}$ fixed. Bearing this in mind, we write down the equations of transformation:

$$\frac{\partial\, {}^4g^{\mu\nu}}{\partial a_{ij}} = -{}^4g^{\mu i}\, {}^4g^{\nu j} + {}^4g^{0\mu}\, {}^4g^{0\nu}\, N^i\, N^j \ ; \qquad (8a)$$

$$\frac{\partial\, {}^4g^{\mu\nu}}{\partial a_{0i}} = -{}^4g^{0\mu}\, {}^4g^{\nu i} - {}^4g^{0\mu}\, {}^4g^{0\nu}\, N^i \ ; \qquad (8b)$$

$$\frac{\partial\, {}^4g^{\mu\nu}}{\partial a_{i0}} = -{}^4g^{\mu i}\, {}^4g^{\nu 0} - {}^4g^{0\mu}\, {}^4g^{0\nu}\, N^i \ ; \qquad (8c)$$

$$\frac{\partial\, {}^4g^{\mu\nu}}{\partial a_{00}} = 2\, {}^4g^{0\mu}\, {}^4g^{0\nu}\, N \ . \qquad (8d)$$

It is straightforward to use Eqs. (7) and (8) to find the non-vacuum ADaM field equations. (Here π^{ij} is the momentum canonical to g_{ij}, defined by Eq. (9c) below. Indices on it and N_i are raised and lowered by the three-dimensional metric, covariant differentiation with respect to which is denoted by a slash, "$|$".)

$$-g^{\frac{1}{2}}\left[\, {}^3R + g^{-1}(\tfrac{1}{2}\pi^2 - \pi^{ij}\pi_{ij})\right] = -2\kappa N^2 g^{\frac{1}{2}} T^{00} \ ; \qquad (9a)$$

$$-\pi^{ij}{}_{|j} = \kappa N g^{\frac{1}{2}}(T^{0i} + N^i T^{00}) \ ; \qquad (9b)$$

$$\partial_t g_{ij} = 2Ng^{-\frac{1}{2}}(\pi_{ij} - \tfrac{1}{2}g_{ij}\pi) + N_{i|j} + N_{j|i} \quad ; \qquad (9c)$$

$$\partial_t \pi^{ij} = -Ng^{\frac{1}{2}}({}^3R^{ij} - \tfrac{1}{2}g^{ij}\,{}^3R) + \tfrac{1}{2}Ng^{-\frac{1}{2}}g^{ij}(\pi^{mn}\pi_{mn} - \tfrac{1}{2}\pi^2)$$

$$-2Ng^{-\frac{1}{2}}(\pi^{im}\pi_m{}^j - \tfrac{1}{2}\pi\pi^{ij}) + g^{\frac{1}{2}}(N^{|ij} - g^{ij}N^{|m}{}_{|m})$$

$$+ (\pi^{ij}N^{|m})_{|m} - N^i{}_{|m}\pi^{mj} - N^j{}_{|m}\pi^{mi}$$

$$+ \kappa Ng^{\frac{1}{2}}(T^{ij} - T^{oo}N^iN^j) \quad . \qquad (9d)$$

I wish to remark on a few features of these equations. First, as we would expect, they do not contain L, since they are simply a linear combination of Eqs. (4). This means they can be used even if a Lagrangian is not available. Second, Eqs. (9) are instructive in understanding even the ADaM vacuum equations, since the particular linear combination used by ADaM is manifest. And third, the equations contain $T^{\mu\nu}$, the contravariant components of the four-dimensional stress-energy tensor. In many situations (e.g., scalar field) one might feel that the covariant components, $T_{\mu\nu}$, are physically more meaningful in a 3 + 1 split, in which case one can rewrite the equations as follows. Using the unit normal to the three-hypersurface, $\eta^\alpha = -N\,{}^4g^{o\alpha}$, one can define a "preferred" energy and momentum density for the matter:

$$\mathcal{E} \equiv \eta^\alpha \eta^\beta \,{}^4T_{\alpha\beta} \,, \qquad (10a)$$

$$\mathcal{P}_i \equiv \eta^\alpha \,{}^4T_{\alpha i} \quad . \qquad (10b)$$

Then the stress tensor in the hypersurface is

$$\mathcal{T}_{ik} = {}^4T_{ik} \quad . \qquad (10c)$$

In terms of these quantities, the relevant parts of Eqs. (9) become

$$- 2\kappa N^2 g^{\frac{1}{2}} T^{oo} = - 2\kappa g^{\frac{1}{2}} \mathcal{E} \quad ; \tag{11a}$$

$$\kappa N g^{\frac{1}{2}}(T^{oi} + N^i T^{oo}) = - \kappa g^{\frac{1}{2}} \mathcal{P}^i \quad ; \tag{11b}$$

$$\kappa N g^{\frac{1}{2}}(T^{ij} - N^i N^j T^{oo}) = \kappa g^{\frac{1}{2}}(N \mathcal{T}^{ij} + N^i \mathcal{P}^j + N^j \mathcal{P}^i) \quad , \tag{11c}$$

where all indices on $\mathcal{P}$ and $\mathcal{T}$ are raised by the three-dimensional metric.

Steps toward a full canonical theory could well begin here. One method would be to specify in advance the motion of the matter in terms of the metric tensor (e.g., homogeneous cosmology), and then to solve the constraint Eqs. (9a,b) by analogy with vacuum ADaM. A more general approach must include a canonical formulation for the fields present in spacetime. In any case, the basic gravitational constraints and dynamical equations will be Eqs. (9).

REFERENCES

1. The best introduction to ADaM is the article by R. Arnowitt, S. Deser, and C. W. Misner in Gravitation, edited by L. Witten (John Wiley and Sons, New York, 1962), Chap. 7.

2. For the Mixmaster Universe, see C. W. Misner, Phys. Rev. Lett. 22, 1071 (1969). For its quantization, see C. W. Misner, "Quantum Cosmology I" (preprint).

3. J. A. Wheeler in Relativity, Groups, and Topology, edited by C. DeWitt and B. DeWitt (Gordon and Breach, New York, 1964) p. 346.

4. L. Landau and E. Lifshitz, The Classical Theory of Fields, (Addison-Wesley, Reading, Massachusetts, 1962), §95.

5. ibid., § 94.

GENERAL RELATIVITY AS A DYNAMICAL SYSTEM ON THE MANIFOLD $\mathcal{Q}$

OF RIEMANNIAN METRICS WHICH COVER DIFFEOMORPHISMS

Arthur E. Fischer[+] and Jerrold E. Marsden[++]

Department of Mathematics
University of California, Berkeley

1. Introduction

In this paper we consider the geometrodynamical formulation of general relativity, due most recently to Arnowitt, Deser, and Misner [2] , DeWitt [3], and Wheeler [8] , from the point of view of manifolds of maps (function spaces) and infinite-dimensional geometry.

Hydrodynamics is approached from this point of view by Arnold [1] and by Ebin-Marsden [4] ; in Fischer-Marsden [5, 6] the function spaces appropriate for a dynamical formulation of general relativity are introduced. We hope that our approach will clarify the basic dynamical structure of the Einstein equations and the associated infinite-dimensional geometry in a spirit analogous to that which has been done in hydrodynamics.

The key to our approach is the group $\mathcal{D} = \mathrm{Diff}(M)$ of smooth (C^∞) diffeomorphisms of a fixed 3-dimensional manifold M. For hydrodynamics one concentrates on $\mathcal{D}_\mu$, the volume preserving diffeomorphisms [4] . For relativity one uses the manifold $\mathcal{Q}$ of Riemannian metrics which cover diffeomorphisms. We begin with a description of $\mathcal{Q}$.

---

†Partially supported by AEC contract AT(04-3)-34.

††Partially supported by Navy contract N00014-69-A-0200-1002
  and NSF contract GP-8257.

2. The Manifold $\mathcal{Q}$ and the Einstein System

Let M be a fixed (no changes in topology) closed (compact without boundary) 3-dimensional oriented smooth manifold, and let

$\mathcal{M}$ = Riem (M) = manifold of smooth Riemannian (positive-definite) metrics on M;

$\mathcal{D}$ = Diff(M) = the group (under composition) of smooth orientation-preserving diffeomorphisms of M; and

$S_2(M)$ = vector space of smooth symmetric 2-covariant tensor fields on M.

Note that $S_2(M)$ is a linear space and that in any decent topology, M is an open convex cone in $S_2(M)$.

Let π : Pos(M) $\to$ M denote the tensor bundle of symmetric positive definite bilinear forms so that $\pi^{-1}(m)$ = space of inner products on T_mM. A Riemannian metric g_n <u>which covers a diffeomorphism</u> $n \in \mathcal{D}$ is a smooth map g_n : M $\to$ Pos(M) such that the following diagram commutes:

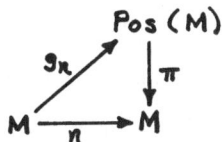

(that is, $\pi \circ g_n = n \in \mathcal{D}$). Thus g_n assigns to each point m $\in$ M an inner product of the tangent space $T_{n(m)}M$. We let $\mathcal{Q}$ denote the manifold of all such maps for all $n \in \mathcal{D}$. $\mathcal{Q}$ is the <u>manifold of Riemannian metrics which cover diffeomorphisms</u>. One can prove that $\mathcal{Q}$ has the structure of a smooth infinite dimensional manifold, cf. [4, § 2]; we shall not require this structure.

There is a natural projection $\bar{\pi}$: $\mathcal{Q} \to \mathcal{D}$ defined by $\bar{\pi}(g_n) = \pi \circ g_n = n \in \mathcal{D}$. Also, if $g_n \in \mathcal{Q}$, observe that $g_n \circ n^{-1} \in \mathcal{M}$ is an "ordinary" Riemannian metric for M. Now $\mathcal{Q}$ is diffeomorphic to $\mathcal{D} \times \mathcal{M}$ by the map

$$\Phi_R : a \to D \times m \; ; \quad g_n \mapsto (n, g_n \circ n^{-1}) ,$$

(Φ_R = right translation) with inverse

$$\Phi_R^{-1} : D \times m \to a \; ; \quad (n, g) \mapsto g \circ n .$$

Thus information on a can be transferred to $D \times m$ and vice-versa via the mapping Φ_R. It is convienient to think of $D \times m$ as a realization of a.

Let $\mathcal{T} = C^\infty (M; \mathbb{R})$ = the vector space of smooth real-valued functions $\xi : M \to \mathbb{R}$ (scalar fields or 0-covariant tensor fields on M).

We will refer to $\mathcal{T}$ as the <u>relativistic time-translation group</u>. Note that the constant functions on M form a subgroup of $\mathcal{T}$ which is isomorphic to R, the <u>classical time-translation group</u>. The manifold $\mathcal{T} \times a \approx \mathcal{T} \times D \times m$ is the proper configuration space for a geometrodynamical formulation of general relativity as we now explain. We will be concerned with the propagation of initial Cauchy data $(g_0, h_0) \in m \times S_2(M)$ off some 3-dimensional hypersurface M of, a yet to be constructed, Ricci-flat (vacuum) space-time V_4. Here $h = \dot{g} = \frac{\partial g}{\partial t}$ is the velocity canonically conjugate to the configuration fields g. As g_t is determined only up to its isometry class, the evolution is determined only up to an arbitrary curve $\mathbf{n}_t \in D$ of diffeomorphisms called <u>the actual shift</u> (with $\mathbf{n}_0 = \mathrm{id}_M$ = e = the identity diffeomorphism); that is, g_t and $(\mathbf{n}_t^{-1})^* g_t$ are isometric evolutions, where

$$(\mathbf{n}_t^{-1})^* \, g_t \, (m) \cdot (Y_m, Z_m) = g_t \circ \mathbf{n}_t^{-1} \, (m) \cdot \left(T_{\mathbf{n}_t^{-1}} (Y_m), T_{\mathbf{n}_t^{-1}} (Z_m) \right), \; Y_m, Z_m \in T_m M,$$

is the "push-forward" of a covariant tensor field. Moreover, one is free to specify on M an arbitrary system of clock rates, or equivalently of clock settings, given as a curve $\xi_t \in \mathcal{T}$ of time functions (<u>the clock settings</u>) with $\xi_0 = \underline{0}$ = the zero function on M (all clocks start at high noon). This arbitrariness or degenency is reflected in the evolution equations as follows:

<u>The Einstein System</u>: <u>Let M be a closed oriented 3-dimensional manifold.</u> <u>Let X_t be an arbitrary time-dependent vector field called the shift vector field</u> <u>and N_t an arbitrary positive scalar field called the lapse function</u>; $N_t(m) > 0$

for all $(t,m) \in \mathbb{R} \times M$. Let g be a given Riemannian metric on M, and let k be a given symmetric 2-covariant tensor field on M such that

$$\delta \left(k - (Tr\ k)g \right) = 0 ,$$

$$\tfrac{1}{2} \left((Tr\ k)^2 - k\cdot k \right) + 2\,R(g) = 0 .$$

The problem is to find a time-dependent metric field g_t on M such that g_t and the supplementary variable

$$k_t = \frac{1}{N_t} \left(\frac{\partial g_t}{\partial t} + L_{X_t} g_t \right) ,$$

satisfy:

   (i) the given initial conditions: $(g_0, k_0) = (g, k)$,

   (ii) the evolution equation

$$\frac{\partial k_t}{\partial t} = S_{g_t}(k_t) - 2N_t\ Ric(g_t) + 2\ Hess(N_t) - L_{X_t} k_t .$$

Our notation is the following:

δk = divergence of k = $(\delta k)_i = k_i{}^j{}_{|j}$ ($|j$ = covariant derivative with respect to the time-dependent metric g),

Trk = Trace k = $g^{ij}k_{ij} = k^i{}_i$,

$k\cdot k$ = dot product for symmetric tensors = $k_{ij}k^{ij}$,

$k*k$ = cross-product for symmetric tensors = $k_{i\ell}k^{\ell}{}_j$,

$S_g(k) = k*k - \tfrac{1}{2}(Trk)k = k_{i\ell}k^{\ell}{}_j - \tfrac{1}{2}(g^{mn}k_{mn})k_{ij}$ = DeWitt spray on $\mathfrak{M}$,

$L_{X_t} g_t$ = Lie derivative of g_t with respect to the time-dependent vector field $X_t = X_{i|j} + X_{j|i}$,

$L_{X_t} k_t$ = Lie derivative of k_t = $X^{\ell}k_{ij|\ell} + k_{i\ell}X^{\ell}{}_{|j} + k_{j\ell}X^{\ell}{}_{|i}$,

$Ric(g_t)$ = Ricci curvature tensor formed from $g_t = R_{ij} =$
$$\Gamma^k_{ij,k} - \Gamma^k_{ki,j} + \Gamma^k_{ij}\Gamma^{\ell}_{k\ell} - \Gamma^{\ell}_{ik}\Gamma^k_{\ell k} ,$$

$R(g_t)$ = Scalar curvature = R^k_k ,

$Hess(N)$ = Hessian of N = double covariant derivative = $N_{|i|j}$.

We now explain how the Einstein system, the lapse function N_t, the shift vector field X_t, and the configuration space $\mathcal{T} \times \mathcal{D} \times \mathcal{M}$ are interrelated (see Fischer-Marsden [5] for more details).

3. The Geometry of the Shift Vector Field

Let $\mathcal{D} = \text{Diff}(M)$, the group of all smooth orientation preserving diffeomorphisms of M. Now $\mathcal{D}$ is a manifold modeled on a Frechet space; (see Ebin-Marsden [4] and related references for the structure of $\mathcal{D}$). The tangent space $T_{\eta}\mathcal{D}$ at a point $\eta \in \mathcal{D}$ is the manifold of smooth maps $X_{\eta} : M \to TM$ which cover η, that is, such that the following diagram commutes:

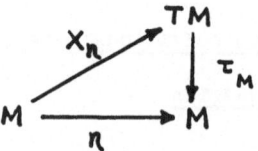

where τ_M denotes the canonical projection of TM to M. To see this let $\eta_t \in \mathcal{D}$ be a curve in $\mathcal{D}$, $\eta_0 = \eta$, so that $\dfrac{d\eta_t}{dt}\Big|_{t=0}$ represents a tangent vector in $T_{\eta}\mathcal{D}$. But for $m \in M$ fixed, $\sigma(t) = \eta_t(m)$ is a curve in M with $\sigma(0) = \eta_t(m)$ and with tangent $\sigma'(0) = \dfrac{d\eta_t}{dt}(m)\Big|_{t=0} \in T_{\eta(m)}M$. Thus $\dfrac{d\eta_t}{dt}$ is a map from M to TM covering η.

We refer to X <u>as a vector field which covers</u> η, so that $T\mathcal{D}$ is the <u>manifold of vector fields covering diffeomorphisms</u>. In particular, $T_e\mathcal{D} = \mathcal{X}(M)$ = the vector space of smooth vector fields on M = the Lie algebra of $\mathcal{D}$. As with the manifold $\mathcal{Q}$, there is a natural projection $\overline{\pi} : T\mathcal{D} \to \mathcal{D}$ defined by $\overline{\tau}(X_{\eta}) = \tau_M \circ X_{\eta} = \eta \in \mathcal{D}$.

Let $\mathcal{R}_{\eta_1} : \mathcal{D} \to \mathcal{D}$ denote right translation by η_1; ($\mathcal{R}_{\eta_1}(\eta) = \eta\eta_1$). Then

$$T\mathcal{R}_{\eta_1} : T\mathcal{D} \to T\mathcal{D} \quad ; \quad X_{\eta} \mapsto X_{\eta} \circ \eta_1 ,$$

so that for $X_{\eta} \in T_{\eta}\mathcal{D}$, $T\mathcal{R}_{\eta^{-1}}(X_{\eta}) = X_{\eta} \circ \eta^{-1} \in T_e\mathcal{D}$ is an "ordinary" vector field on M, called <u>the pull-back</u> of X_{η} <u>by right translation.</u>

Now let $X_t : M \to TM$ be a time-dependent vector field on M. Then the flow η_t of X_t with η_0 = identity is a smooth curve in $\mathcal{D}$ (as X_t is time-dependent, η_t is not a one-parameter subgroup of $\mathcal{D}$) which satisfies

$$\frac{d\eta_t}{dt} = X_t \circ \eta_t \; , \quad \text{or} \quad \frac{d\eta_t}{dt} \circ \eta_t^{-1} = X_t \; .$$

Conversely, given a smooth curve $\eta_t \in \mathcal{D}$ with η_0 = identity, $\frac{d\eta_t}{dt} \circ \eta_t^{-1} = X_{\eta_t} \circ \eta_t^{-1} = X_t$ is a time-dependent vector field which generates η_t as its flow.

Thus in the Einstein system, if one gives the <u>shift vector field</u> X_t, then the <u>actual shift</u> of M is its flow $\eta_t \in \mathcal{D}$, a curve in $\mathcal{D}$. Equivalently one may specify the actual shift $\eta_t \in \mathcal{D}$ and compute the shift vector field as above. It is because of the presence of the shift vector field that the group must be included in the configuration space.

The relationship between the Lie derivative terms and the shift vector field can be explained geometrically as follows. Suppose that for $\bar{N}_t = 1$, $\bar{X}_t = 0$, $(\bar{g}_t, \bar{k}_t) \in \mathfrak{M} \times S_2(M)$ is a solution to the Einstein system with initial conditions $(\bar{g}_0, \bar{k}_0)$; that is,

$$\frac{\partial \bar{g}_t}{\partial t} = \bar{k}_t \; ,$$

$$\frac{\partial \bar{k}_t}{\partial t} = S_{g_t}(\bar{k}_t) - 2 \, \text{Ric} \, (\bar{g}_t) \; .$$

Now let X_t be an arbitrary shift vector field with flow η_t, η_0 = identity. Then $(g_t, k_t) = ((\eta_t^{-1})^* \bar{g}_t, (\eta_t^{-1})^* \bar{k}_t)$ are solutions to the evolution equations with $N_t = 1$, X_t = given shift vector field, and the same initial data as before. This follows by a direct verification:

$$\frac{\partial g_t}{\partial t} = \frac{\partial (\eta_t^{-1})^* \bar{g}_t}{\partial t}$$

$$= (\eta_t^{-1})^* \frac{\partial \bar{g}_t}{\partial t} - L_{X_t} ((\eta_t^{-1})^* \bar{g}_t)$$

$$= k_t - L_{X_t} \, g_t \; ,$$

where we have used the fact that

$$\frac{d}{dt} \left(\eta_t^{-1} \right)^* g = -L_{X_t} \left(\eta_t^{-1} \right)^* g \; ; \; \text{see } [7] \, , \; p. \; 32 \, .$$

Similarly,

$$\frac{\partial k_t}{\partial t} = \frac{\partial \left(\eta_t^{-1} \right)^* \bar{k}_t}{\partial t}$$

$$= \left(\eta_t^{-1} \right)^* \frac{\partial \bar{k}_t}{\partial t} - L_{X_t} \left(\eta_t^{-1} \right)^* \bar{k}_t$$

$$= S_{g_t} (k_t) \quad -2 Ric (g_t) - L_{X_t} k_t$$

since $S_{\bar{g}} (\bar{k})$ and $Ric (\bar{g})$ are tensors and hence commute with $\left(\eta_t^{-1} \right)^*$; that is, $\left(\eta_t^{-1} \right)^* (Ric (\bar{g})) = Ric \left(\left(\eta_t^{-1} \right)^* \bar{g} \right) = Ric(g)$.

The significance of this result may be clarified as follows: Besides the realization of $\mathcal{Q}$ as $\mathcal{D} \times \mathcal{M}$ by "right translations," there is a realization of $\mathcal{Q}$ as $\mathcal{D} \times \mathcal{M}$ by "left translations" defined as follows:

$$\Phi_L : \mathcal{Q} \rightarrow \mathcal{D} \times \mathcal{M} \; ; \quad g_\eta \mapsto \left(\eta^{-1} \right)^* \left(g_\eta \cdot \eta^{-1} \right) \, .$$

These two realizations of $\mathcal{Q}$ are entirely analogous to the two realizations of TSO(3) for the rigid body into body and space coordinates respectively; see Arnold [1] . Thus the introduction of a shift may be viewed merely as shifting from body to space coordinates by use of the coordinate change η_t .

4. The Lapse Function and the Intrinsic Shift Vector Field

To discuss the lapse we assume that the shift vector field $X_t = 0$. (They can be handled simultaneously by using the semi-direct product on $\mathcal{T} \times \mathcal{D}$.) If we choose the lapse $N_t = 1$, then the evolution of g is parameterized by a canonical evolution parameter, the proper time τ. But suppose that g is a solution of the Einstein system for an arbitrary lapse N. One constructs a space-time on $\mathbb{R} \times M$ in a tubular neighborhood of M by the Lorentz metric (in coordinates)

$$g_{\mu\nu} dx^\mu dx^\nu = -N^2 dt^2 + g_{ij} dx^i dx^j .$$

The proper time function $\tau(t,m) = \tau_t(m) = \tau(t,x^k)$ (in this tubular neighborhood of M) is then just the time coordinate in Gaussian normal coordinates $(\tau(t,x^k), \bar{x}^i(t,x^k))$, where $\bar{x}^i(t,x^k)$ is the space part of the Gaussian coordinates. To find the relation between the lapse N_t and τ_t, we consider the transformation of $g_{\mu\nu}$ to Gaussian normal coordinates; writing out $\bar{g}^{00} = g^{\mu\nu} \frac{\partial \tau}{\partial x^\mu} \frac{\partial \tau}{\partial x^\nu}$ yields

$$-1 = -\frac{1}{N^2}\left(\frac{\partial \tau}{\partial t}\right)^2 + g^{k\ell} \frac{\partial \tau}{\partial x^k} \frac{\partial \tau}{\partial x^\ell} ,$$

which is solved for N_t to give

$$N_t = \frac{d\tau_t}{dt} \frac{1}{\sqrt{1 + \|\text{grad } \tau_t\|^2}} ,$$

where $\|\text{grad } t\|^2 = g^{k\ell} \frac{d\tau}{dx^k} \frac{d\tau}{dx^\ell}$ is computed with respect to the inverse $g^{k\ell}$ of the time-dependent metric $g_{k\ell}$ ($= {}^4g^{ij}$ since the shift is zero). The factor $\frac{1}{\sqrt{1 + \|\text{grad}\tau\|^2}}$ takes into account the fact that in general the lapse depends on space coordinates and therefore pushes up the hypersurface M through $\mathbb{R} \times M$ unevenly.

The single first order partial differential equation for τ

$$\left(\frac{d\tau}{dt}\right)^2 - N^2 g^{k\ell} \frac{d\tau}{dx^k} \frac{d\tau}{dx^\ell} = N^2$$

can be reduced to a system of eight first-order ordinary differential equations by the Cauchy method of characteristics. Of course this system of ordinary differential equations is just the system of geodesic equations of the Lorentz metric $g_{\mu\nu}$ (for unit timelike geodesics) in Hamiltonian form. If we choose on the non-characteristeric hypersurface $t = 0$ the initial condition: $\tau(0,m) = 0$ (corresponding to geodesics normal to $t = 0$), then we are assured of a unique $\tau(t,m)$ that satisfies the above equation with the initial condition $\tau(0,m) = 0$. Note that $\frac{d\tau}{dt} = N$ on this initial hypersurface.

The condition

$$\bar{g}^{oi} = -\frac{1}{N^2} \frac{\partial \bar{x}^i}{\partial t} \frac{\partial \tau}{\partial t} + \frac{\partial \bar{x}^i}{\partial x^n}\left(g^{mn} \frac{\partial \tau}{\partial x^m}\right) = 0$$

gives an equation for the space part $\bar{x}^1(t, x^k)$ of the Gaussian normal coordinate system,

$$\frac{\partial \bar{x}^i}{\partial t}(t, x^k) = \frac{N(t, x^k)}{\sqrt{1 + \|grad\tau\|^2}} \; \frac{\partial \bar{x}^i}{\partial x^n}(t, x^k)\left(g^{mn}(t, x^k)\frac{\partial \tau}{\partial x^m}(t, x^k)\right) \quad .$$

Set $Y_t = -\dfrac{N(t, x^k)}{\sqrt{1 + \|grad\tau\|^2}} \; grad\tau$; then the above equation can be written as

$$\frac{d\varphi_G}{dt} = -D\varphi_G \cdot (Y) \;,$$

where φ_G is the spatial part of the Gaussian normal coordinates and $D\varphi_G$ is, in coordinates, the Jacobian matrix of φ_G . But the identity

$$\frac{d}{dt}\left(f_t^{-1} \cdot f_t\right) = \frac{df_t^{-1}}{dt}\cdot f_t + Df_t^{-1}\cdot\frac{df_t}{dt} = \frac{df_t^{-1}}{dt}\cdot f_t + Df_t^{-1}\cdot Y_t \cdot f_t = 0$$

then shows that this equation is solved by $\varphi_G = f_t^{-1}$ if f_t is the flow of Y_t. We call Y_t the <u>intrinsic shift of the lapse</u> since it describes the "tilting" of the Gaussian normal coordinates due to the space dependence of the lapse function. The above argument shows that the partial differential equation for the space part of the Gaussian normal coordinate system can be solved by an ordinary differential equation, namely finding the flow of the intrinsic shift. Finally, the inverse to the contravariant metric

$$\bar{g}^{ij}\left(\tau(t, x^k), \bar{x}^i(t, x^k)\right) = \frac{\partial \bar{x}^i}{\partial x^m}(t, x^k)\frac{\partial \bar{x}^j}{\partial x^n}(t, x^k) g^{mn}(t, x^k) - \frac{1}{N^2}\frac{\partial \bar{x}^i}{\partial t}(t, x^k)\frac{\partial \bar{x}^j}{\partial t}(t, x^k)$$

$$= \frac{\partial \bar{x}^i}{\partial x^m}(t, x^k)\frac{\partial \bar{x}^j}{\partial x^n}(t, x^k)\left(g^{mn}(t, x^k) - \frac{1}{1 + \|grad\tau\|^2}g^{m\ell}(t, x^k)\frac{\partial \tau}{\partial x^\ell}(t, x^k) g^{nr}(t, x^k)\right.$$

$$\left. \cdot \frac{\partial \tau}{\partial x^r}(t, x^k)\right)$$

solves the evolution equations with N = 1 (and the same initial data) if $g_{ij}(t,x^k)$ solves the Einstein equations with an arbitrary N. Writing g^{-1} for the contravariant components of g, the above equation can be written intrinsically as

$$\bar{g}^{-1}\left(\tau(t,m),\, \varphi_G(t,m)\right) = D\varphi_G(t,m) \otimes D\varphi_G(t,m)\left(g^{-1}(t,m) - \frac{\text{grad}\,\tau(t,m)}{\sqrt{1+\|\text{grad}\,\tau\|^2}} \otimes \frac{\text{grad}\,\tau(t,m)}{\sqrt{1+\|\text{grad}\,\tau\|^2}}\right).$$

Our prescription shows how, given a solution to the Einstein equation with an arbitrary N, to find the solution to the Einstein equations with N = 1 and the same initial data by solving ordinary differential equations only. A similiar prescription is available to go from solutions for N = 1 to solutions for arbitrary N; see [5]. To take into account the lapse function we introduce the <u>relativistic time translation group</u> $\mathcal{T} = C^\infty(M;\mathbb{R})$ (a group under pointwise addition of functions). As $\mathcal{T}$ is a vector space, $T\mathcal{T} = \mathcal{T} \times \mathcal{T}$. For a given lapse N_t and a solution g_t to Einstein's equations with this lapse, we construct a curve $\tau_t \in \mathcal{T}$ such that

$$\left(\frac{d\tau}{dt}\right)^2 - N^2\,\|\text{grad}\,\tau\|^2 = N^2$$

and $\tau_0 = 0$. Thus to find the curve in $\mathcal{T}$ corresponding to a given lapse N we must first solve Einstein's equations with this particular lapse.

In the case that N depends only on the time coordinate, then τ_t and N_t are simply related by $\tau_t = \int_0^t N_\lambda \, d\lambda$. Moreover, if $(\bar{g}_t, \bar{k}_t)$ is a solution to the Einstein system with initial conditions $(\bar{g}_0, \bar{k}_0)$ and lapse $\bar{N}_t = 1$, then the solution with $N_t = f(t)$ $\left(\text{and } X_t = 0\right)$ and the same initial conditions is just the reparameterized curve $(g_t, k_t) = (\bar{g}_{\tau(t)}, \bar{k}_{\tau(t)})$. This is easily seen, as

$$\frac{\partial g_t}{\partial t} = \frac{\partial \bar{g}_{\tau(t)}}{\partial t} = \frac{\partial \bar{g}_{\tau(t)}}{\partial \tau}\frac{d\tau(t)}{dt} = N_t\,\bar{k}_{\tau(t)} = N_t\,k_t$$

and

$$\frac{\partial k_t}{\partial t} = \frac{\partial \bar{k}_{\tau(t)}}{\partial t} = \frac{\partial \bar{k}_{\tau(t)}}{\partial \tau}\frac{d\tau(t)}{dt} = N_t\left(S_{\bar{g}_{\tau(t)}}(\bar{k}_{\tau(t)}) - 2\,\text{Ric}(\bar{g}_{\tau(t)})\right)$$

$$= N_t\,S_{g_t}(k_t) - 2N_t\,\text{Ric}(g_t).$$

5. <u>The Einstein Lagrangian on</u> $\mathcal{T} \times \mathcal{Q} \approx \mathcal{T} \times \mathcal{D} \times \mathcal{M}$

Since $\mathcal{M}$ is an open convex cone in $S_2(M)$, $T\mathcal{M} = \mathcal{M} \times S_2(M)$. On $\mathcal{M}$ we define the <u>DeWitt metric</u> $\mathcal{G}$ (see DeWitt [3] , and Fischer-Marsden [5]) by

$$\mathcal{G}_g : T_g \mathcal{M} \times T_g \mathcal{M} = S_2(M) \times S_2(M) \rightarrow \mathbb{R}$$

$$\mathcal{G}_g (h_1, h_2) = \int_M \left((\mathrm{Tr}\, h_1)(\mathrm{Tr}\, h_2) - h_1 \cdot h_2 \right) \mu_g ,$$

where μ_g is the volume element associated with the metric g (in coordinates $\mu_g = \sqrt{\det\ g}\ dx^1 \wedge dx^2 \wedge dx^3$). $\mathcal{G}$ is a non-degenerate but weak metric on $\mathcal{M}$; here weak means that the map $\mathcal{G}_g^\# : T_g\mathcal{M} \rightarrow T_g^\#\mathcal{M}$, defined by $\mathcal{G}_g^\#(h_1) \cdot h_2 = \mathcal{G}_g (h_1, h_2)$ is an injection, by the non-degeneracy, but is not an isomorphism.

We now introduce a potential V $: \mathcal{M} \rightarrow \mathbb{R}$ defined by

$$V(g) = 2 \int_M R(g) \mu_g$$

(twice the integrated scalar curvature). If on $T\mathcal{M}$ we consider the Lagrangian

$$L = T - V : T\mathcal{M} = \mathcal{M} \times S_2(M) \rightarrow \mathbb{R} ,$$

defined by $\qquad L(g,h) = \frac{1}{2} \mathcal{G}_g (h,h) - V(g)$,

then a computation shows that Lagrange's equations give the Einstein system with lapse $N_t = 1$ and shift $X_t = 0$.

The DeWitt metric $\mathcal{G}$ on $\mathcal{M}$ is extended to $\mathcal{D} \times \mathcal{M} \approx \mathcal{Q}$ by defining on each fiber $T_{(\eta, g)} (\mathcal{D} \times \mathcal{M}) = T_\eta \mathcal{D} \times S_2(M)$

$$\mathcal{G}(\eta, g) : \left(T_\eta \mathcal{D} \times S_2(M) \right) \times \left(T_\eta \mathcal{D} \times S_2(M) \right) \rightarrow \mathbb{R}$$

$$\mathcal{G}(\eta, g) \left((X_{\eta_1}, h_1), (X_{\eta_2}, h_2) \right) = \mathcal{G}_g (h_1 + L_{X_{\eta_1} \circ \eta_1^{-1}}\, g,\ h_2 + L_{X_{\eta_2} \circ \eta_2^{-1}}\, g).$$

The Lagrangian L on $T\mathcal{M}$ is now extended to a Lagrangian on $T(\mathcal{D} \times \mathcal{M})$ by

$$\overline{L} : T(\mathcal{D} \times \mathcal{M}) = T\mathcal{D} \times \mathcal{M} \times S_2(M) \rightarrow \mathbb{R}$$

$$\overline{L} (X_\eta, g, h) = L (g,\ h + L_{X_\eta \circ \eta^{-1}}\, g)$$

$$= \frac{1}{2} \mathcal{A}_g (h + L_{X_n \circ n^{-1}} g, \quad h + L_{X_n \circ n^{-1}} g) - V(g) .$$

Note that the factor $\mathcal{D}$ is now essential as X_n is explicitly involved in L.

Now $\mathcal{A}$ is a degenerate metric on $\mathcal{D} \times \mathcal{M}$ since if

$$\mathcal{A}(n,g) \left(h + L_{X_n \circ n^{-1}} g, \quad k + L_{Y_n \circ n^{-1}} g \right) = 0 \quad \text{for all} (Y_n, k) \in T_n \mathcal{D} \times S_2(M),$$

then

$$h + L_{X_n \circ n^{-1}} g = 0 ,$$

but h and X_n need not be zero independently. This degeneracy has the effect of introducing some ambiguity into the equations of motion. However, the degeneracy of $\mathcal{A}$ is such that we are free to specify a curve of diffeomorphisms $n_t \in \mathcal{D}$; thus the ambiguity in the equations of motion is completely removed by the specification of the shift vector field X_t.

Using $\overline{L} : T(\mathcal{D} \times \mathcal{M}) \to \mathbb{R}$, we construct on $T(\mathcal{T} \times \mathcal{D} \times \mathcal{M})$ the

Einstein Lagrangian

$$L_E : T(\mathcal{T} \times \mathcal{D} \times \mathcal{M})$$

defined by

$$L_E(\xi, N, X_n, g, h) = \int_M N \left\{ \left(\frac{h + L_{X_n \circ n^{-1}} g}{N} \right) \cdot \left(\frac{h + L_{X_n \circ n^{-1}} g}{N} \right) - \left[Tr \left(\frac{h + L_{X_n \circ n^{-1}} g}{N} \right) \right]^2 \right\} \mu_g$$

$$- 2 \int_M N \ R(g) \mu_g .$$

L_E now picks up a degeneracy in the $\mathcal{T}$ direction, as well as in the $\mathcal{D}$ direction, allowing for the arbitrary specification of N_t as well as X_t. However, once N_t and X_t are specified, the degeneracy of L_E is completely removed and the evolution equations are well-defined. A computation then shows that Lagrange's equations in the "non-degenerate direction", together with the arbitrarily specified lapse function N_t and shift vector field X_t, are the Einstein equations of evolution (see [5] for details).

BIBLIOGRAPHY

1   V. Arnold, _Sur la géométrie differentielle des groupes de Lie de dimension infinie et ses applications a l'hydrodynamique des fluids parfaits_, Ann. Inst. Genoble 16 (1) (1966), 319-361.

2   R. Arnowitt, S. Deser and C. W. Misner, _The Dynamics of General Relativity_, in : _Gravitation; An Introduction to Current Research_, ed. L. Witten, Wiley, New York (1962).

3   B. DeWitt, _Quantam Theory of Gravity. I. The Canonical Theory_, Phys. Rev. 160 (1967), 1113-1148.

4   D. Ebin and J. Marsden, _Groups of Diffeomorphisms and the Motion of an Incompressible Fluid_, Ann. of Math. 92 (1970) 102-163.

5   A. Fischer and J. Marsden, _The Einstein Equations of Evolution - A Geometric Approach_, to appear.

6   A. Fischer and J. Marsden, _The Geometry of the Einstein Evolution Equations_, to appear.

7   S. Kobayashi and K. Nomizu, _Foundations of Differential Geometry_, Interscience (1963).

8   J. A. Wheeler, _Geometrodynamics and the Issue of the Final State_, in _Relativity, Groups and Topology_, ed. C. DeWitt and B. DeWitt, Gordon and Breach, New York (1964).

Lecture Notes in Physics

Selected Issues from
Lecture Notes in Mathematics

Beschaffenheit der Manuskripte
Die Manuskripte werden photomechanisch vervielfältigt; sie müssen daher in sauberer Schreibmaschinenschrift mit ausreichend großer Type geschrieben sein. Handschriftliche Formeln bitte nur mit schwarzer Tusche eintragen. Notwendige Korrekturen sind bei dem bereits geschriebenen Text entweder durch Überkleben des alten Textes vorzunehmen oder aber müssen die zu korrigierenden Stellen mit weißem Korrekturlack abgedeckt werden. Die reproduktionsfähigen Abbildungen (in Originalgröße) sollen in den Text eingeklebt werden. Falls das Manuskript oder Teile desselben neu geschrieben werden müssen, ist der Verlag bereit, dem Autor bei Erscheinen seines Bandes einen angemessenen Betrag zu zahlen. Die Autoren erhalten 50 Freiexemplare.

Zur Erreichung eines möglichst optimalen Reproduktionsergebnisses ist es erwünscht, daß bei der vorgesehenen Verkleinerung der Manuskripte der Text auf einer Seite in der Breite möglichst 18 cm und in der Höhe 26,5 cm nicht überschreitet. Entsprechende Satzspiegelvordrucke werden vom Verlag gern auf Anforderung zur Verfügung gestellt.

Manuskripte, in englischer, deutscher oder französischer Sprache abgefaßt, sind einzureichen bei: Springer-Verlag, 6900 Heidelberg, Postfach 1780.

Cette série a pour but de donner des informations rapides, de niveau élevé, sur des développements récents en physique, aussi bien dans la recherche que dans l'enseignement supérieur. On prévoit de publier.

1. des versions préliminaires de travaux originaux et de monographies

2. des cours spéciaux portant sur un domaine nouveau ou sur des aspects nouveaux de domaines classiques

3. des rapports de séminaires

4. des conférences faites lors de congrès ou de colloques

En outre il est prévu de publier dans cette série, si la demande le justifie, des rapports de séminaires et des cours multicopiés ailleurs mais déjà épuisés.

Dans l'intérêt d'une diffusion rapide, les contributions auront souvent un caractère provisoire; le cas échéant, les démonstrations ne seront données que dans les grandes lignes. Les travaux présentés pourront également paraître ailleurs. Une réserve suffisante d'exemplaires sera toujours disponible. En permettant aux personnes intéressées d'être informées plus rapidement, les éditeurs Springer espèrent, par cette série de «prépublications», rendre d'appréciables services aux instituts de physique. Les annonces dans les revues spécialisées, les inscriptions aux catalogues et les copyrights rendront plus facile aux bibliothèques la tâche de réunir une documentation complète.

Présentation des manuscrits
Les manuscrits, étant reproduits par procédé photomécanique, doivent être soigneusement dactylographiés type assez grand. Il est recommandé d'écrire à l'encre de Chine noire les formules non dactylographiées. Les corrections nécessaires doivent être effectuées soit par collage du nouveau texte sur l'ancien soit en recouvrant les endroits à corriger par du vernis correcteur blanc. Les illustrations; en dimension originale, préparées pour reproduction sont à insérer dans le texte. S'il s'avère nécessaire d'écrire de nouveau le manuscrit, soit complètement, soit en partie, la maison d'édition se déclare prête à verser à l'auteur, lors de la parution du volume, le montant des frais correspondants. Les auteurs reçoivent 50 exemplaires gratuits.

Pour obtenir une reproduction optimale il est désirable que le texte dactylographié sur une page ne dépasse pas 26,5 cm en hauteur et 18 cm en largeur. Sur demande la maison d'edition met à la disposition des auteurs du papier spécialement préparé.

Les manuscrits en anglais, allemand ou français peuvent être adressés à Springer-Verlag, 6900 Heidelberg, Postfach 1780.